Chris Defonseka

Polymeric Coating Systems for Artificial Leather

Also of Interest

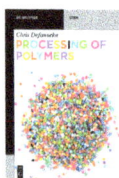

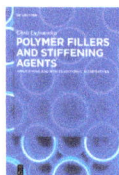

Chris Defonseka

Polymeric Coating Systems for Artificial Leather

Standard and Latest Technologies

DE GRUYTER

Author
Chris Defonsekat
Toronto
Canada
defonsekachris@rogers.com

ISBN 978-3-11-071653-5
e-ISBN (PDF) 978-3-11-071654-2
e-ISBN (EPUB) 978-3-11-071663-4

Library of Congress Control Number: 2021943183

Bibliographic information published by the Deutsche Nationalbibliothek
The Deutsche Nationalbibliothek lists this publication in the Deutsche Nationalbibliografie;
detailed bibliographic data are available on the Internet at http://dnb.dnb.de.

© 2022 Walter de Gruyter GmbH, Berlin/Boston
Cover image: DGolbay/iStock/Getty Images Plus
Typesetting: Integra Software Services Pvt. Ltd.
Printing and binding: CPI books GmbH, Leck

www.degruyter.com

Preface

From ancient times, the use of animal skins for wear and other necessities has been common practice. Even today, the use of the skins of particular animals is highly prized, and after special curing processes, they are termed *natural leather*. Some luxury automobile manufacturers may even go to the extent of having their own farms for particular animal's skins.

As the world populations increased, the demand for leather for many applications such as upholstery, footwear, travel bags and hand bags escalated. Since natural leather was very expensive and in short supply, industrial manufacturers came out with cloths coated with plastics which was termed as *Rexine*, the quality of which was poor due to limited technical knowledge. Some of the challenges the manufacturers had to face were colour migration, surface cracking, brittleness when exposed to sunlight and poor surface finishes.

However, the researchers and developers in this field had to increase their activity to come up with better components to produce better quality artificial leather as worldwide protests grew against the use of natural animal skins. Polyvinyl chloride (PVC) polymer coatings on cotton cloth produced acceptable artificial leathers as an alternative to natural leather. Aesthetically pleasing surface finishes could be achieved, but the perennial problem of cracking and brittleness remained after exposure to sunlight for some time.

With rapid developing technology, polyurethane polymers were found to give softer and more supple artificial leather, though more expensive and PVC leather being less expensive and being of reasonable qualities remains popular to this day. This presentation takes the manufacture of artificial leather one step further in introducing silicone-coated leather as the newest technology, and coupled with the author's new concepts of using both bamboo-woven and bamboo-knitted fabrics and others coupled with recommendations for the use of non-traditional fillers, stiffening, softening agents, graphene and the latest ultraviolet countering plasticizers should be both challenging and exciting.

It is hoped that this book will provide the necessary impetus to researchers, polymer coating system developers and artificial leather manufacturers to meet the challenges as experienced by the end users.

Chris Defonseka

https://doi.org/10.1515/9783110716542-202

Acknowledgements

This book is a well-researched presentation of the methodologies for manufacturing artificial leather, which includes standard and latest technologies. The author has been working closely with manufacturers of this commodity which is an essential product needed by consumers, footwear industry and the automobile industry among others. The research and inputs provided by manufacturers in North America, Europe and China and the technical information provided by various organizations are invaluable and the author appreciates their cooperation.

The new concepts and coating technologies presented will no doubt be of great interest to readers of this book. The author acknowledges the valuable support and guidance offered especially by Dr Christene Smith (acquisitions editor), Anna Bernhard (contents editor) and her staff at De Gruyter to make this highly technical book possible.

Chris Defonseka

https://doi.org/10.1515/9783110716542-203

Contents

Chapter 9
A small manufacturing plant for artificial leather for an entrepreneur —— 105

Chapter 1
Introduction

1.1 What is artificial leather?

From ancient times, people used well-cured skins of animals for clothing, rugs, tents, footwear and in other applications. As population grew over the years and these sources diminished, they began to look for alternative materials.

To this day, natural leather is highly valued and considered number one, like preference for natural wood products against synthetic products such as furniture. It is of no doubt that well-cured natural leather from animals has a class of its own but technology has advanced so much that imitation or synthetic artificial leather products are equally good and are much cheaper than natural leather products. In certain applications like upholstery in high-end automobiles, certain manufacturers have their own farms producing these special 'natural' leathers from animal hides.

Artificial leather, also called synthetic or imitation leather, is a material made as a substitute for natural leather for upholstery, clothing, footwear, handbags, protective wear and tarpaulins, and the applications are endless. These polymeric coated materials give a very close finish to natural leather and are very cost-effective. Unlike in natural leather, the surface finishes that can be achieved with artificial leather are amazing. In addition to the standard embossed patterns in glossy, matt or fluffy (suede) surfaces and in almost in any colour, there are top-end finishes like imitation of crocodile skin, snake skin, silver and gold colours which are in great demand. Another aspect of artificial leather category is transparent polymeric coatings on synthetic films, paper and printed cloths for clothing. This presentation will also introduce new concepts like the use of *bamboo cloth* (woven and knitted), *handloom cloth* and woven cloths from natural fibres with specially formulated polymeric coatings, some of which will be transparent coatings.

Imitation leather, another name for artificial leather, as the name suggests, is a material that looks like real leather. There are two basic methods of manufacturing artificial leather. In *direct coating*, a fabric is coated with a polymeric mix directly onto the surface of the fabric. This could be either a solid or foamed mix. For high-end foamed artificial leather, the preferred method would be *indirect coating*, where a solid polymeric mix is first coated onto an embossed release paper and then foam mixes are coated and laminated with a fabric. Since the production of natural leather is expensive due to wastage costs incurred in the cutting process due to the irregular contours and to preserve the inherent grain structures and natural markings, constant attempts have been taking place to develop materials that mimic the surface and feel of natural leather. Advancing technologies have made it possible to produce materials very close to natural leather so much, so that some may find it difficult to distinguish between the two.

https://doi.org/10.1515/9783110716542-001

Visually, manufacturers have succeeded very well in imitating a natural leather surface. However, artificial leather is unable to match some technical properties of leather like *haptics, breathability, water vapour permeability* and *toughness*. Therefore, imitation leather remains as an inexpensive alternative to genuine leather. However, there are also areas that withstand constant contact with sunlight and water, which would damage real leather. Also, in the medical field, in applications like dentist's chairs, massage beds, examination and treatment beds, artificial leather will be more durable to withstand the use of disinfectants, which are mostly solvent based and would dissolve the finish of leather and finally make it brittle. Artificial leather is also widely used for protective wear such as safety aprons, gloves and industrial boots. These materials will not have an embossed surface and instead have a plain surface and probably with a glossy finish for easy cleaning. In keeping with industrial safety codes, the colour could be yellow with polymeric coatings on both sides with the fabric sandwiched in between.

Other categories with polymeric coatings, although they cannot be strictly called artificial leather but certainly belonging to it as a branched 'family', would be paper-coated products, coated or embossed synthetic films and other substrates with polymeric coatings to suit particular end applications.

1.2 Visual differentiating

Differentiating between genuine leather and artificial leather is not always easy, even for experts. Some may recommend even checking them in a laboratory. If these materials are already used, for example, in upholstery, then it is difficult as only the surface is available for scrutiny. On the other hand, if these two materials are available in their normal forms, it is easier to identify each. The author suggests, as a simple test, to examine the back of both materials and the genuine leather will have no 'backing material', while artificial leather will have one.

Artificial leather will have a textile backing, generally cotton or synthetic and they will be dyed the same colour as the coating to improve quality, mainly aesthetic value. Genuine leather will have a slightly fibrous back but then some imitation leathers also have this effect. If the material consists of a foam layer that is thick enough, one could squeeze the foam and observe the 'bounce back', whereas in genuine leather, it is solid. The grain pattern in genuine leather is irregular, whereas the embossed surface patterns of artificial leather are neat and uniform.

When synthetic leather is cut, the cutting edge is often smooth and clean, whereas when leather is cut it has an almost 'lint-like' effect. Natural leather has a fibrous structure that can be seen under a magnifying glass. The grain side has a dense fibre layer which becomes more fibrous towards the centre. In the case of artificial leather, the top layer is very dense and without fibres and will have a fabric backing. In some cases, artificial leather can also take the form of a fabric coated on both sides (protective wear) and also a pure sheet of solid or foam.

1.2.1 Behavioural patterns

Imitation leather melts at high temperatures and burns well, while natural leather will only glow and solidify without catching fire. Burnt artificial leather will smell like burnt plastic, whereas leather smells like burnt hair.

1.2.2 Stretchability

Synthetic leather is often thinner and stretches more when heated. If the materials are glued on hard surfaces, for example, steering wheels and table tops, they cannot be tested and an inexperienced person will not be able to recognize the difference. Some may argue that artificial leather is colder than leather but there are exceptions. Good imitations with smooth or rough surfaces can feel warmer than some leathers. With modern-day technology, it is possible to copy the natural haptic of a thick aniline leather.

1.2.3 Breathability

A distinctive feature between the two is the breathability aspect. Even a heavily coated leather has more breathability than unperforated artificial leather. In rubber boots, one would sweat more than in waterproof boots made of leather. When one sits on a synthetic leather surface, after sometime, one would sweat more quickly than on real leather. This is due to the porous factor of leather. However, artificial leather can be micro-perforated to make it more vapour-permeable.

1.2.4 Other characteristics

Some may argue that rubbing artificial leather would produce more creaking noises than leather. This may well depend on the quality of the materials. High-end materials of both kinds of patent leather and top-end artificial leather like glossy plain surfaces will produce 'creaking' noises, while matt finishes like suede will not.

If artificial leather is damaged due to age and long use, it is usually clear immediately. Synthetic leather often breaks and the fabric backing underneath becomes visible. On the other hand, natural leather suffers damage slowly and will have a different appearance.

Sometimes, both types of leathers are used in one application. A good example is the upholstery in automobiles, where identical looking natural leather and artificial leather are used. Usually, the contact surfaces such as seats, back rest and armrests are upholstered in natural leather, while the rest is done in artificial leather. For an untrained eye, it will be almost impossible to tell the difference.

It is often said that genuine leather can be recognized by its smell. Good well-cured natural leather has a characteristic 'leather' smell which is very hard to achieve in artificial leather. Also, different types of genuine leather may have different smells. Most manufacturers of artificial leather use 'leather bouquets' (perfumes compatible with surface coating) to match natural leather smells and also to give off a nice fragrance. If odour-giving stabilizers like amines or others are used in polyvinyl chloride or polyurethane coatings, then the use of a 'leather bouquet' to mask the odour becomes essential. These fragrances can be single or a combination of more than one to get the maximum effect. A lingering soothing 'smell' over a long period of time is most. Here, some caution may be required to ensure that these fragrances used to have a neutral effect on people to counter possible allergies but in general terms this has not been an issue.

1.3 Different surface finishes

In keeping with the needs for different types of artificial leather for market like consumer, automobile, furniture, industrial, footwear, fashionwear, decorative, entertainment, travel goods, safety clothing or other, each application will need different qualities with different surface finishes like matt, glossy, suede, psychedelic, embossed, plain, wood grain and animal skin imitations.

Bibliography

[1]   The Leather Dictionary. www.leather-dictionary.com.
[2]   Defonseka, Chris. Defonseka Technology Consultants.

Chapter 2
Types of artificial leather

2.1 Brief history

Over the years, many different types of artificial leathers have been developed, with probably the USA, the UK and Germany being the pioneers. Today, several countries are into the manufacture of these synthetic or imitation leathers for natural leather, which are expensive and in short supply. Since there is an expanding market demand for these products, constant research is being carried out to find more and more versatile polymeric coatings, not necessarily dependent on traditional ones, to meet newer challenges in diverse applications.

Probably, one of the earliest examples was *Presstoff* invented in the nineteenth century in Germany. It was made of specially layered and treated paper pulp. It was in great demand during the Second World War in place of natural leather which was in short supply and rationed. Presstoff could be used in most applications where natural leather was used, except in applications where flexing and moisture, for example in footwear, would delaminate the layers due to loss of cohesion.

Another early example was *Rexine*, a leather cloth produced in the UK by Rexine Ltd. in Manchester. It was made of cloth coated with a mixture of nitrocellulose, camphor oil, alcohol and pigment and embossed to look like leather. It was used as a bookbinding material, upholstery in motor vehicles and railway carriages. A big advantage was its cost being only around a quarter that of natural leather.

Du Pont in the USA came up with artificial leather called *Poromerics*, made from a plastic coating on a fibrous base layer, typically a polyester cloth. The term 'poromerics' was coined by them as a derivative of the terms *porous* and *polymeric*. They introduced this first poromeric material under the brand name *Corfam* at a shoe show in Chicago in 1963. After marketing it to shoe manufacturers, Du Pont sold their rights to accompany in Poland in 1971.

Leatherette, a combination of a fabric and a plastic coating also came on the market among some others. This was made using a backing fabric made of natural or synthetic fibre, which was then coated with a soft polyvinyl chloride (PVC) layer and carried a simple surface design. This product was mainly used in book binding with different colours to suit the end applications.

With rapidly developing technologies, many new types of artificial leathers are being made using both cotton, synthetic and other fabrics and newly developed polymeric coatings or combinations of them. Among these, polyurethanes (PUs), PVC and now silicones are the most popular choices.

https://doi.org/10.1515/9783110716542-002

2.2 Brand names

There are many brand names and different qualities of artificial leather on the market today. Probably, the most high-end qualities required are for automobiles, footwear, furniture and the fashion industries.

The following are some of the brand names:
- Clarino – artificial leather made by Japanese company
- Fabrikoid – Du Pont: cotton cloth coated with nitrocellulose
- Lorica – manufactured by Lorica Sud, an Italian company
- Naugahyde – an American product by Uniroyal
- Rexine – made by British company
- Kirza – an early Russian product consisting of cotton cloth, latex and rosin
- Pinatex – made from pineapple leaves
- Kantala – Sri Lankan company specializing in leather made from Hana plant
- Faux leather – PVC/PU coated fabrics
- Vegan leather – non-animal artificial leather
- Bluesil leather – silicone coated artificial leather
- Skai artificial leather

2.2.1 What is faux leather?

Faux leather is one of the several names given to artificial, synthetic or imitation leather. These names often identify themselves with specific end uses of synthetic leathers such as faux leather (sofa, chair and headboard upholstery), leatherette (auto upholstery, clothing) and koskin (consumer goods). In all cases, the basic polymeric coatings are PVC and PU.

Vinyl (PVC) synthetic leathers have been manufactured as early as 1940, initially, for shoes, automobile interiors and furniture upholstery. In early 1950s, Du Pont and other chemical companies began to develop PU products and the end-users found them better than PVC coated leathers due to its extra softness and flexibility. However, each one is better for certain applications than the other. Also, costs were also a factor with the PUs costing more.

Since PUs were more flexible and 'breathable', they were the preferred material for clothing, auto interiors, luxury furniture and so on, especially where it came in direct contact with skin.

This presentation will be mainly based on artificial leather made with polymeric coatings of PVC, PU and silicone (Si). Also discussed will be other speciality polymeric coatings suited for special applications.

Figure 2.1 shows some of the types of artificial leather currently being manufactured.

Figure 2.1: Types of artificial leather.

2.3 Polymeric coated artificial leather

In this chapter, we will discuss three main types of artificial leather which will cover most end applications. The big volume consumptions and requirements are in the automobile, footwear, furniture, travel bags and fashion wear sectors. It is interesting to note that as challenges arise for the end applications of artificial leather as the markets become more sophisticated and demanding, the advent of newer materials such as silicones, graphene, high-heat resistance plasticizers, self-heat controlling coatings and other things compatible with polymeric coating systems is providing manufacturers with great solutions. In some cases, these newer polymeric systems are enabling end-users to use these materials in newer cost-effective applications.

2.3.1 PVC coated artificial leather

The basic concept of PVC coated leather is a polymeric mixture of PVC polymer, coated on either woven or knitted fabrics. These fabrics can be either cotton or synthetic and, generally, cotton fabrics are used for *direct coating*, where the PVC mixture is directly applied on to the surface of the fabric. The first coat will be a base coat on which a manufacturer will build the final required thickness by one or two more coats and then a thin final coat, which will be embossed as desired.

If a more softer and flexible leather is desired, a manufacturer will apply a polymeric mixture containing a blowing agent as the middle layer and if the colour of the base fabric being used match the colour of the coatings, the final product will give much better aesthetic appearance.

In order to produce really soft-textured foamed artificial leather which is very close or almost identical to natural leather, a manufacturer would use an *indirect coating method* using a knitted fabric and an embossed release paper on to which the first layer is cast/coated. Then, one or two foam coats are added and finally the knitted fabric is laminated on to this foam coat. This way one could produce some really high-end artificial leather. Naturally, there would be a significant cost difference between the direct coated and indirect coated materials.

These surfaces will have different finishes like matt, glossy, plain and psychedelic with different colours as desired. Since the plasticizers used in a polymeric mix is susceptible to heat and has a tendency to slowly evaporate, making the leather less flexible after long use, manufacturers may use a very thin transparent coating on the final surface to prevent this action on the surface. A good example is the interior of an automobile. After long exposure to sunlight, one may notice the beginning of a thin vapour coating on the windows (fogging) due to some of the plasticizer evaporation. Since, standard artificial leather is not porous and could cause sweating, modern day technologies allow the manufacturers to counter this by making more 'breathable' leather.

2.3.2 Polyurethane coated artificial leather

These types of leather give much better results than the ones with PVC. The qualities that can be achieved by coating fabrics with PU is such that the top-end products will need an expert to identify them from real well-cured natural leather.

The chemicals that go to make up a polymeric PU mixture can be varied to obtain different final textures. Where PU coatings are concerned, a manufacturer probably would use an indirect coating method over a direct coating method to obtain maximum quality. However, one would have to exercise more caution when dealing with PU chemicals which are both flammable and toxic than with PVC components. To avoid this factor to some extent, a manufacturer has the choice of purchasing fully mixed PU coatings ready for use. One drawback would be that the range would be limited as each pre-mixed purchase could produce only one quality.

Here, a manufacturer may use either knitted rayon or nylon fabrics, coloured the same as the coatings to maximize aesthetic values. Pleasing colours and designs can be achieved and PUs being more flexible than PVC will be ideal for fashion wears also.

2.3.3 Silicone coated artificial leather

For many years, artificial leather has been accepted as either PVC or PU coated fabrics. Both of these have their own strengths and weaknesses but they have been found to be a really good substitute for natural genuine leather. Both qualities have been found adequate for end applications covering a very wide range of applications. PVC leathers has been long recognized as the go-to workhorse uses, especially in public spaces, while PU are recognized for their soft hand unique soft and very supple feeling and versatile design options. Irrespective of the cost differences, both can be used for any application depending on the affordability and end purpose.

However, a new product – silicone coated artificial leather – entering the market has created a new excitement with newer possibilities. Although silicones have been around for many years and used for electronics, medical, consumer products, building construction and others, now silicone-coated textiles in the form of artificial leathers is even a closer imitation to genuine leather but still costs less. From 100% silicone coatings to silicone hybrids and top coats, silicone without a doubt is the most exciting polymer for polymeric coatings for textiles. Elkem Silicones markets them under their brand name *Bluesil*.

2.3.4 Faux/vegan leather

Faux leather, also referred to as Leatherette or vegan leather, is often considered a lower-end cost alternative to genuine leather. PVC faux leather is made by coating a fabric with a PVC plastisol containing a PVC polymer, plasticizer, stabilizers and a filler. The addition of a dye will give a desired colour and an embossing, a surface finish.

PU faux leather is made by direct coating on to a fabric which can be cotton, polyester, rayon or nylon. Also, these PU coatings can be applied by an indirect coating/laminating method. Different finishes can be achieved by adding colour and an embossing pattern to imitate genuine leather.

These cost-effective formulated and made products may not last long and thus are used for lower-end applications.

2.3.5 Genuine leather

Commonly called natural leather is made from organic materials, typically bovine hides. In the fashion wear trade, these hides can extend to other animals. Over the years, these products have been on the decline due to increasing protests giving rise to increased market demands for artificial leather.

The natural collagen fibres of these hides are intricately intertwined, providing superior durability over man-made woven or knitted products. There are numerous variations of genuine leather with different types of finishes depending on the curing and finishing processes. All genuine leathers are inherently stronger than artificial leathers.

2.3.6 Skai artificial leather

Skai artificial leathers, made with PVC and special surface coatings offer a high-quality range of synthetic leathers in place of genuine leather. Skai imitation leathers with specialized surfaces are used internationally in a variety of applications in the interior designs of high-quality living, medical sector and also for outdoor applications. These leathers effortlessly meet all requirements with regard to fire protection (B1 certified) and for comfort and provide a good substitute for genuine leather. Since they provide real comfort in upholstery, furniture fronts, durability among other properties, they are accepted as genuine quality products.

According to reports, the contract sector places high demands on Skai upholstery fabrics. Two of their special brands – Skai *Parotoga NF* and Skai *Palma NF* – meet the very strict standards required by the building codes, especially for flame-retardant properties, are in demand for applications in hotels, cruise ships or for seating in restaurants and stadiums and so on. Some of the other important properties of these fabrics are *durability* and *abrasion*.

When creating their pieces, furniture manufacturers and designers as well as interior designers find real inspiration with attractive designs and trendy colours on offer from these leathers. According to the manufacturers of Skai leathers, designers, time and again make very irresistible, emotionally impactful living spaces, whether in homes, hotels, outdoor living or luxury resorts.

The product characteristics of certain Skai-coated fabrics such as *Surface Plus* makes them ideal for the contract sector such as nursing, retirement homes, hospitals and doctor's surgical needs as they meet the highest medical requirements. The 'Plus' stands for outstanding surface resistance to oil, grease and sweaty substances, with the surfaces being anti-bacterial and anti-microbial proof. These specially coated polymeric surface coatings are resistant to disinfectants, blood and urine, and meets the German medical devices act.

Skai cool colours *Venezia* and *Neptune* offer innovative upholstery materials for outdoor applications. These special materials are high-quality upholstery fabrics when used outdoors the surfaces will significantly reduce the warming of the upholstery material due to sunlight. Specially selected cool colour pigments will reflect about 80% of infrared radiation and keep the surfaces cooler. The Skai Neptune outdoor material is available in many summer colours and will withstand all weather conditions. This extremely robust material is lightfast and waterproof and has many other properties.

A speciality artificial leather *Laif-VYP* in the Skai leather's range includes a breathable leather permeable to air. This innovative material belongs to the next generation breathable, water vapour permeable polymeric fabric materials. Whether in a deceptively realistic genuine leather look, metallic shine or textile look and feel manufacturers of Shai leathers claim their products are real alternatives to natural genuine leather.

2.3.7 Leo Vinyls – PVC coated artificial leather

Leo Vinyls Ltd., an Indian company, reports that they produce high-class artificial leathers in the form of PVC coatings on cotton cloth. The formulations are varied to suit the different qualities of leathers produced to cover a wide range of end applications as follows:

- Automotive
- Footwear
- Luggage
- Stationery
- Furniture upholstery
- Sports goods
- Garments
- Furnishings

2.3.8 POLYFABS artificial leather

Premier Polyfilm Ltd. India produces both PVC and PU coated fabrics with basic standard specifications of 2.0 m width and up to an overall thickness of 2.0 mm. Their range of products covers a range as follows:

- Automotive
- Belts
- Shoes
- Bags
- Furnishings

Their products can be specially made to custom requirements of properties to include any of the following or combinations:

- Colour fastness
- High flexing
- Cold resistance (cracking)
- Abrasion resistant
- Flammability
- Anti-fogging
- Mildew resistant
- Weather resistant
- Stain resistant
- Salt water resistant
- Alcohol resistant

There are many brands and qualities of artificial leather being produced by a wide range of industrial countries and it is not possible to include all in this presentation. It is hoped that the information provided in this section will give the reader an idea of the different types of artificial leather being produced to meet the needs of an ever expanding end uses. Later, in Chapter 6, the more advanced coating systems will be presented.

Bibliography

[1]   Skai Artificial Leather. www.skai.com/en/interior/artifical-leather.
[2]   "What is PVC Leather?" Bridgesi.com –www.bridgesi.com/fashion/what-is-pvc-leather.html.
[3]   Burdett, Jennifer. "Types of Artificial Leather." www.leaf.ty/13719926/whatisveganleather.
[4]   Artificial Leather- Imitation Leather. www.LEATHERDICTIONARY.COM.
[5]   Elkem. "Bluesil": Silicone Coated Artificial Leather. www.elkem.com.

Chapter 3
Understanding dyes, pigments and colouring

3.1 Colour basics

All plastic products need colour, especially so for artificial leather. Although in this case the quality of the final material takes precedence, the aesthetic values that colours provide are an essential feature, especially from a marketing angle. In a very sophisticated market, just any colour would not do for most applications. For example, different colours, shades and tints are required for footwear, automobile applications, upholstery and furniture applications, fashion wear and so on. Most polymeric mixtures are easy to colour but to obtain the maximum value, it is best that the colourants used, in whatever form be it powders, liquids, pastes or other things, are compatible with the polymeric mixture. If the mixture contains more than one polymer, the colourants must be compatible with both or in some cases additive compatibilizers can be used.

All dyes, pigments and colouring stem from a generally accepted colouring system consisting of three basic colours, that are yellow, red and blue, but the author would like to add two more colours – white and black – as shown below:
– White
– Black
– Yellow
– Blue
– Red

Many other colours can be obtained by mixing these basic ones.

 For example yellow + blue = green
 Black + white = grey
 Red + white = pink
 Blue + red = brown

In addition to these 'straight' colours by mixing more than two, several other colours/shades can be obtained. Silver and gold colours stand alone as speciality colours, and in the manufacture of artificial leather, these are important especially for the footwear and fashions industries.

 Pigments and dyes are distinctly different types of colourants. A pigment is a finely divided solid which is essentially insoluble in a polymer application medium. Pigments are incorporated by a dispersion process into the polymer while it is in a liquid phase, and after the polymer solidifies, the dispersed pigment particles are retained physically within the solid polymer matrix.

 In contrast, a dye dissolves in a polymer application medium and is usually retained as a result of an affinity between individual dye molecules and molecules of

https://doi.org/10.1515/9783110716542-003

the polymer. Pigments are generally preferred to dyes for colouring of plastics mainly because of their superior colour fastness properties and especially resistance to colour migration.

Although synthetic materials are man-made, it is not unreasonable to think that they are in harmony with nature. Colours are more, and in the manufacture of artificial leather, they play a big role when it comes to its end applications. A good example is its use in automobile application.

Colourants can be ranked according to their tone, pigment class, colour index, opacity, light-fastness, tinting strength and physiological/chemical properties. Unlike pigments, colourants are combined with a polymer matrix and generally soluble in plastics. Manufacturers of colourants try to maintain equilibria between natural and synthetic materials, other than the standard ones with many manufacturers offering custom-made colours, matrices and properties to suit any application. With advances in masterbatch colouring, additive masterbatches are available with additives incorporated for ultraviolet (UV), flame retardants and other things.

3.2 Theory of colours

All colours originate from sunlight and a spectrum of light consists of violet, indigo, blue, green, yellow, orange, red (VIBGYOR). Using colours, one could set a mood, attract attention, create an atmosphere or make a statement. Colours can be used to energize or cool down and by choosing the right colour scheme, one could create an ambience of elegance, warmth or tranquillity. A colour or a combination of colours can be a most powerful design element in a final product if one knows how to use them effectively.

Sunlight gives many thousands of colours but our eyes can perceive only a few as shown in a spectrum. Colours affect us in numerous ways, both mentally and physically. For example, a strong red colour may raise one's blood pressure, whereas a blue colour will have a calming effect. Being able to offer aesthetically pleasing artificial leather or coated fabrics especially for the automobile, furniture, footwear and fashion industries can lead to spectacular results.

3.3 The colour wheel

The colour wheel or colour circle is a base tool for combining colours. According to reports, the first colour wheel was designed by Sir Isaac Newton. Over the years, many variations of this basic design have been made but the most common version is the original colours of 12 colours based on red, yellow and blue (RYB). In general, several colour combinations are considered as pleasing which are called *colour harmonies* or *colour chords* and they consist of two or more colours with a fixed relationship in the colour wheel. *Colour impact* is designed to dynamically create a colour wheel to match your base colour. Colours are broadly categorized into *primary, secondary* and *tertiary* colours.

3.4 Primary colours

In the RYB colour model, the primary colours are red, yellow and blue. I would like to include black and white making it five colours instead of three. Primary colours mean that these colours cannot be achieved by mixing any other colours. With this base, it is possible to obtain any other colour by mixing among them.

3.5 Secondary colours

The three secondary colours, green, orange and purple are obtained by mixing two of the primary colours as follows:
Green: yellow + blue
Orange: red + yellow
Purple: red + blue

3.6 Tertiary colours

An additional six tertiary colours are obtained by mixing primary and secondary colours.

3.7 Warm and cool colours

This is an important aspect where artificial leather is concerned when applicable, for example, in automobile and the furniture industries. Warm colours are vivid and energetic and tend to 'expand' in space, whereas cool colours exude an impression of calm and create a soothing effect. White, black and grey are considered to be neutral colours.

3.8 Tints, shades and tones

Tints, shades and tones are often used incorrectly, although they are simple colour concepts. If a colour is made lighter, it is called a *tint*. If black is added, the darker version is called a *shade*. If, grey is added, the result is a *different tone*.

3.9 Colour harmonies

Complimentary colour schemes are colours that are opposite to each other on the colour wheel, for example red and green. The high contrasts of complimentary colours create a vibrant look, especially if used in full concentration. However, if it is not properly used, it may give a 'jarring' effect. Complimentary colour schemes are difficult to use in large amounts but work well if you want something to stand out.

Analogous are often found in nature and are harmonious and pleasing to the eye. One must ensure that there is sufficient contrast if choosing an analogous colour scheme. Choose one colour to dominate, a second colour in support and a third along with black, white or grey as an accent.

Triatic colour schemes use colours that are evenly spaced around the colour wheel. These colour schemes tend to be quite vibrant, even if one uses pale or unsaturated versions of colour hues. To achieve a triatic harmony successfully, the colours should be carefully balanced, for example, one colour to dominate and two others to accent. Split complimentary colour schemes are variations of the complimentary colour schemes. In addition to the base colour, two other colours adjacent to its complement are used. This colour scheme has the same strong visual contrast as the complimentary colour but has less tension.

3.10 Masterbatches

Masterbatches are concentrated colourants in a polymer medium to colour or obtain special effects. If the medium matches or at least is compatible with the polymer to be coloured, maximum effects can be achieved. It is also possible to use more than one masterbatch but usual practice is the use of a single one. A masterbatch is used in low concentration as an additive into a larger batch during compounding. Most thermoplastic polymers are used in particulate structure because this will permit considerable less pollution during processing and easier assimilation. Most manufacturers prefer to supply masterbatches as concentrated granules or as liquids.

Some additive masterbatches may contain mineral fillers, for example mineral fillers for modifications of resins such as polyethylene or polypropylene, with significant ones being calcium carbonate and barium sulphate. Previously, most fillers were valued for their economic value but appreciation for mineral fillers has increased as their influence on the functionality and physical properties of plastic products has become apparent. Using fillers, the density, weight, anti-block and damping properties of a polymer can be changed easily.

3.11 Concentrated masterbatches

The most common method used by processors is masterbatches consisting of concentrated pigments dispersed into a polymer carrier resin. During processing the masterbatch, the masterbatch is mixed with natural resin and can be compounded with other ingredients. In the case of artificial leather, the pigment concentrate can be mixed with the filler content and a small amount of a plasticizer and milled on a two-roll or three-roll mill to obtain a very fine paste before being mixed with the polymeric mix. The percentage of this 'colour batch' will depend on the final colour desired but generally, small. Most processors of artificial leather will have large batches of these colourants ready as stock in a wide variety to meet the standard range of coloured products. This will help in cutting down production time.

Depending on the polymer matrix being used for coating, there are formulae to calculate the amount of colour needed as a percentage. If more than one polymer is used in a polymeric mix, compatibility becomes important and an artificial leather manufacturer will work closely with a supplier/laboratory assistance to get the maximum results. When dealing with high-end applications, for example automobiles, colour performance will also take precedence.

3.12 Universal masterbatches

Universal masterbatches are offered by some companies, one of which is RTP Company (USA) marketing under the brand name *Unicolor* as an innovative solution for colouring multiple polymers including engineered polymers. These products are universal masterbatches with extremely low let-down ratios, typically 1–2%, depending on the resin and final colour desired and they do not require drying before processing.

These products are a great solution for processors that want one colour products in a wide range even from different polymers and in the manufacture of artificial leather where the basic polymers would be PVC, PUR and silicones.

Some of the main colours offered are white, magnolia, beige, terracotta, brown, yellow, amber, nectarine, red, raspberry, fluorescent green, green blue, royal blue and light grey. Variations of these colours are also available and also special 'customer' colours. For artificial leathers, three of the basic important properties when choosing colours are colour fastness, high resistance to colour migration and ease of miscibility/assimilation in the polymeric mix. The 'reds' always pose problems with colour migration tendency and also colour fastness. Colours carry numbers, and the higher the number, the better.

Colour matching (custom orders) techniques are now highly advanced and processors of polymeric coatings will have no problems in obtaining what they require. Processors who have their own in-house laboratories will have the advantage of trying out colours in 'mini' trial runs and also experimenting with different polymeric coating systems to improve their standard product lines. With rapid advancement in technology the market sees many new products and these can be tried out and used to improve product lines.

3.13 Colourants

As mentioned earlier, colourants can be divided into two categories, pigments and dyes. Pigments are organic or inorganic materials that are practically incompatible with polymeric materials and must be dispersed in another medium, for example in a filler, and then mixed with the polymer batch. Pigments come in several forms such as powders, granules, liquids, solid concentrates and masterbatches tailored by compounders.

Because dyes dissolve in the resin, there are no visible particles and they will not affect the transparency of the mix. When selecting colourants some of the important aspects are as follows:
- *The colour index* is a classification for dyes and pigments
- *Heat resistance* – check the highest processing temperature and exposure time.
- *Light fastness* – performance rating between 1 (low performance) and 8 (high performance)
- *Weather fastness* – most supplier will provide this information
- *Migration* – dyes and pigments can migrate to the surface. This is called *blooming* or *bleeding*.
- *Abrasion* – many inorganic pigments are very abrasive, thereby causing damage to processing equipment during long-time use.
- *Chalking* – If too much pigment is used, a surface degrades during weathering.

Note: A supplier will provide excellent information on the correct pigment or dye to use. However, processing conditions, temperatures and others may also be a factor. Perhaps a processor may want to carry out an *ageing test* on a coated cloth or have this done outside.

3.14 Pigments – important properties

Pigments are insoluble organic or inorganic particles added to a polymer base or plastisol to give a specific colour to the product. Pigments that are organic in nature are hard to disperse and tend to form agglomerates (clumps). These can cause lumps, spots, specks and other small surface defects and, in this case, the surface coating of artificial leather. On the other hand, inorganic pigments like metal oxides, sulphides, carbon black and so on get more easily dispersed in polymer resins.

Amongst the inorganic pigments, titanium dioxide is the most widely used pigment in the plastics industry, particularly used to obtain white colour.

In polymeric coatings for artificial leather, the colouring or a colouring system used must have some basic important properties to achieve quality, especially products that are to be used outdoors. Some of them are as follows:

3.14.1 Weatherability/ageing

Exposure to sunlight and some artificial lights can have adverse effects on coloured plastics products, more so, on soft polymeric coated surfaces like artificial leather. Excess heat or constant exposure to sunlight for long periods will cause the coloured surfaces to somewhat 'fade'. This condition can also cause the plasticizer in a PVC coating to slowly evaporate over a long period of time and make a surface loose its flexibility. In the case of solvent-based polyurethane coatings, the same process can occur. Some manufacturers of artificial leather may apply a very thin transparent surface coating which will be effective from solvents or plasticizer evaporation but not the effects of sunlight.

To assess the weathering patterns of polymeric coated surfaces, a manufacturer will have to have laboratory tests done in relation to the climate and weather patterns of the region where these materials will be used. However, this is not easy as different regions may have different weather patterns. Selecting UV stabilizer systems and colour fastness additives, especially for the surface coats must be done carefully.

3.14.2 Light fastness

In general terms, colour fastness means the degree of resistance to light of a colouring product applied to a plastic product, in this case the polymeric coatings applied. Since the intermediate layers are protected by a final top coat, colour fastness will apply to these coats. Colour fastness pigment grades may differ for indoor and outdoor applications.

While straight colours are more prone to loss of colour or fading as such, with reds being the most difficult, pastel shades or very light colours may withstand light action better. Pigment selection for indoor use depends on:
- Polymer types
- Concentration of pigment used
- Presence of titanium dioxide (accelerates fading)
- Required level of light fastness (specs.)
- Type of usage

Pigment performance can also be influenced by:
- Type of surface coating
- Processing heat conditions
- Heat stabilization system used

Generally, inorganic pigments will exhibit superior light fastness and when selecting pigments for polymeric coatings a supplier's guidance will be helpful and in most cases will have laboratory facilities to make and supply custom made pigment grades.

3.14.3 Transparency

Transparent coats can be easily obtained by using a normal formula without titanium dioxide or fillers like calcium carbonate in PVC formulations. These coats are often used by manufacturers of artificial leather to protect the top coats. Some of the problems have been losing flexibility due to loss of plasticizer or solvents where they have been used. Also, these coatings can be scratch resistant.

Iron oxide pigments can be opaque or transparent. The transparent variety is an important group of inorganic pigments as they are used for metallic finishes, where high levels of transparency gives an attractive finish. Moreover, their weatherability resistance improves the weatherability of pigments with which they can be combined. This is known as a synergistic effect.

Pigment groups are mainly categorized into:
- Organic pigments
- Inorganic pigments
- Carbon black
- White pigments
- Special effects pigments
- Aluminium pigments
- Other pigments

3.14.4 Pigment pastes for PVC coatings

Some manufacturers of artificial leather may prefer to buy ready-made pigment pastes to colour their polymer coatings. These products, both organic and inorganic pigments, are dispersed in plasticizers in such a way that can be utilized particularly for production of PVC synthetic leather and will add the required colour to polymeric coatings. There are many reputed suppliers and according to one such supplier – BASPAR LIA chemical company – their range of pigment pastes – CH-600 series, as presented in Table 3.1 – has been tested for heat resistance, ageing and colour fastness and if they are stored in properly sealed containers, they can be stored for at least 1 year at 50 °C.

Table 3.1: Range of pigment pastes by BASPAR LIA chemical company (modified by author; specifications courtesy of BASPAR Chemical Co).

Product name	Product code	Solid, %	Heat resistance, °C	Light resistance
White Lia paste	CH-600	55	300	8
Yellow Lia paste	CH-610	55	180	7
Yellow Lia paste	CH-612	55	180	7
Yellow Lia paste	CH-613	45	180	6
Orange Lia paste	CH-636D	60	200	7–8
Red Lia paste	CH-633 C	25	240	7–8
Brown Lia paste	CH-650 T	60	300	8
Brown Lia paste	CH-651A	60	300	8
Brown Lia paste	CH-655D	60	300	8
Blue Lia paste	CH-661	30	300	8
Black Lia paste	CH-670	20	300	8
Green Lia paste	CH-681	30	300	8

Note: Light resistance index 8 is considered high, 5 average and 3 low. Heat resistance factor is suitable for processing temperatures.

3.14.5 Colourants for polyurethane coatings

Among the range of artificial leathers, polyurethane coated ones are considered as high-end products, although more expensive than the hereto popular PVC coated ones. The polyurethane products' markets have been expanding rapidly over the years with products like flexible foam, semi-flexible foams, upholstery and automotive applications, bedding and furniture, integral skin products, PU footwear, insulation foams and such with PU artificial leather, a huge market.

To make these products aesthetically pleasing, polyurethane colourants of very high quality is manufactured by many companies using pigments and other polymeric raw materials. One such company – Everlight Chemical – produces and offers reactive polymeric colourants specially designed for polyurethanes under the brand name – *Evertint*.

These could be used in various polyurethane applications as foam, elastomers, adhesives, synthetic leather and casting PU. Unlike the traditional PU colour pastes composed of pigment dispersion, Evertint contains multiple hydroxyl groups to react with the isocyanate group of PU and will form chemical bonding into the PU polymer matrix. With its reactive nature, Evertint provides strong colour fastness and solvent resistance. In keeping with environmental concerns, these PU colourants use 'green chemistry' and all production processes do not include any harmful solvents such as dimethylformamide.

When adding colour to polyurethanes, a manufacturer of artificial leather may want to consider the use of *SO-Strong* liquid urethane colourants made by Smooth-On Inc., USA, which can be used to create a variety of colour effects for a wide range of polymeric coating systems. For glowing fluorescent colour effects, for example for clothing apparel for the entertainment industry or fashionwear, their grades of Ignite colourants can be used. Colour tints are also possible but the grade or grades to be used should be compatible with the PU coating systems. Their range of SO-Strong, UVO and Ignite colourants are highly concentrated, offering excellent dispersion and consistent colour. A very small amount will colour a proportionately large amount of liquid urethane or epoxy materials. The supplier recommends a loading range of 0.01–3.00% of total system weight. Loading in excess may cause cure inhibition or oozing.

Another company which produces and offers suitable colourants for polyurethanes is Colortek India Ltd. in collaboration with Morten Technology Ltd. UK. Colortek colourants are suitable for colouring all types of polyurethanes like flexible slab-stock foams, automotive applications, integral skins, PU footwear, synthetic coatings and others. These polyurethane colourants are manufactured with high-quality pigments and other polymeric raw materials. The data given below explains some of the main features of these products:

- Low-viscosity pastes
- Phthalate (plasticizer)-free colours for various PU applications
- Special non-yellowing white colour paste with UV resistance
- Melamine-based pastes for fire retardant and acoustics applications
- High jet-black colour based on polyols and phthalate free for PU applications
- Polyol based, economical high-strength colourants, especially for footwear
- Finely dispersed pigment pastes, non-fading
- Compatible with both polyether and polyester polyol-based PU systems
- High-viscosity pastes for blending in polyol to get maximum yields
- Customized colouring solutions as per individual needs

Thus, manufacturers of artificial leather with different PU polymeric systems will have a wide choice range to suit most applications and the added advantage of ordering speciality colour pastes for customized orders.

3.14.6 Colourants for silicone coatings

When artificial leather is made with silicone coatings, the surfaces give excellent combinations of softness and 'rubbery' feeling with toughness. This may be attributed to the fact that silicones can be called polymeric material and belonging to the rubber family as well, which in turn with rubber being classified as a polymer also. The normal sequence is for the supporting coatings to be of lighter colour and the top coat or the 'skin' to have a full colour as desired. Unlike colouring polymeric PVC or polyurethane coatings, one may think more of tints or shades to obtain the full aesthetic value of silicones finishes.

When colouring silicones, there are options one of which is to use Silc Pig or Silc Pig Electric pigments which are highly concentrated and made by Smooth-On Inc. silicone products. Because they are concentrated dispersions, the use of small amounts will go a long way and the recommended loading range is between 0.01% and 3.00%. The availability of a small in-house laboratory is invaluable to enable a manufacturer to try out different colourants and especially when tints are used. The best practice would be to mix in the colourant directly into the polymeric mix and mix thoroughly until a homogenous coloured mix is obtained before using it for coating.

3.15 Dyes for colouring fabrics

Dyes may be defined as substances that when applied to a substrate provides a change in colour from its original colour. This process at least temporarily alters any crystal structures of the coloured products. Such substances with considerable colouring capacity are widely used in textiles, pharmaceuticals, food, cosmetics, plastics, photographic and paper industries. For textiles, the dyes can be either synthetic or from natural sources like plants, roots or other. The use of coloured backing fabrics to match the colour of the coatings in artificial leather will enhance the aesthetic values.

The dyes can adhere to compatible surfaces by solution, by forming covalent bonds or complexes with salts or metals, by physical absorption or by mechanical retention. Naturally, for different fabrics like cotton, nylon, rayon, polyester, bamboo cloth and all fabrics that can be used in the manufacture of artificial leather, a processor of dyed fabrics will select compatible dyes, and for high-end product needs, the best dyes irrespective of the cost factor.

The fabric dyeing process uses a lot of water and it is estimated that many tons of dyes are also wasted during a dyeing and finishing process due to effluents. Unfortunately, these dyes escape conventional waste-water treatment processes and persist as an environmental concern due to their resistant to high stability to light, temperature, water, detergents, chemicals, soap and bleach. Synthetic origins and complex aromatic structures of these make them more resistant to bio-degradation.

According to statistics based on production volumes, *azo dyes* are the largest group of colourants, comprising 60–70% of all organic dyes produced in the world. The success of these azo dyes is due to their ease of use and cost-effectiveness for synthesis as compared to natural dyes and also their great structural diversity, high molar extinction coefficient and medium to high fastness properties in relation to light as well as wetness. However, some azo dyes may show toxic effects.

3.15.1 The dyeing process in brief

Dyeing methods have not changed much with time. Basically, water is used to clean, dye and apply auxiliary chemicals to the fabrics and also to rinse the treated fabrics. The basic dyeing process involves three steps *preparation, dyeing* and *finishing* as follows:

Preparation is the step in which unwanted impurities are removed from the fabric before dyeing. This can be carried out by cleaning aqueous alkaline substances and detergents. Many fabrics are bleached with hydrogen peroxide or chloride containing compounds in order to remove their natural colour and if the fabric is to be sold as white and not dyed, optical brightening agents are added.

Dyeing is the aqueous application of colour to the textile substrate, mainly using synthetic organic dyes and frequently at elevated temperatures and pressures in some of the steps. During this step, the dyes and chemical aids such as surfactants, acids, electrolytes, levelling agents and others are applied to the textile to get a uniform depth of colour and achieve the degree of colour fastness required for the end application. This process involves diffusion of the dye into the liquid phase followed by absorption into the outer surface of the fibres of the fabric and diffusion into the inner surfaces of the fibres also. Depending on the colourfastness requirements of the end use of the final coloured product, the correct suitable dyes must be chosen, also keeping in mind that PVC is processed at temperatures around 220 °C but polyurethane coatings at lower temperatures. Colour fastness is a very important factor, for example coloured fabrics for swimsuits must not bleed in water and artificial leather or coloured fabrics in automobile applications should not fade.

For manufacturing artificial leather, the backing fabrics have to be in long lengths of rolls and preferably without joints. In continuous processing, heat and steam are applied to long rolls of fabrics as they pass through a series of concentrated chemicals solutions. The fabric retains the greater part of the chemicals, while rinsing removes the excess chemicals.

Finishing involves treatments with chemical compounds aimed at improving the quality of the fabric. Some of these are permanent press treatments, water proofing, softening, anti-static protection, soil resistance, stain resistance, microbial/fungus protection, abrasion resistance and minimize shrinkage.

3.16 Custom colours for polymeric coatings

If processing straight polymer resins such as PVC, PUR or silicones, one may say that it is easy to colour and process. However, when the coatings include foamed polymers (best grades), the ability to obtain colour uniformity, prevention of migration, stability of colour and so on becomes a little more difficult. Many factors can affect colour, including resins, additives, textures, finishes, UV action, processing conditions. However, with modern technology, these can be easily overcome, especially if one works closely with suppliers or colour experts. Modern day colour instruments will also help.

The surface finish required by customers will also have some effect on colour and when special colours are needed, the costs of colour matching will also be a factor. Some common finishes/textures can be identified as glossy, matt, textured, grain, suede or psychedelic.

3.17 Preparation of colour batches

Manufacturers of artificial leather have the choice of using dyes or pigments to colour the polymeric coatings. The use of pigments probably is the preferred one for coatings, while dyes are used for colouring the substrates as the backing like fabrics, paper and films. In a fully self-contained manufacturing operation, it is best to purchase the individual components of all raw materials needed and do the polymeric coating mixtures on the production floor rather than purchase custom-made mixtures according to one's specifications. This way, a manufacturer will have the freedom of mixing different formulations as and when needed to suit each production run. Of course, for this one must have the technical know-how with an in-house polymer chemist and at least a small lab, which will be greatly helpful in colour-matching for custom needs or producing a manufacturer's own range of colours. Alternatively, a manufacturer could use outside facilities or work in conjunction with suppliers who will be only too willing to help. If patents are involved or special colours are formulated with proprietary rights, then one could work with special agreements.

If dyes are used, they can be directly mixed into the polymeric mix and due to the stirring action, air bubbles will be generated. If the coating is to be a solid one, for example in a direct coating process, these air bubbles must to removed, probably using a vacuum action, while these bubbles will be helpful if the coating is a foamed one. Needless to say that all dyes or pigments used must be compatible with the polymeric mix.

When using pigments, a manufacturer could mix them with the filler, for example, calcium carbonate and obtain a thoroughly homogenous mix on either a two-roll or three-roll mill. Some may want to mix several batches of different colours as stock and draw from them as and when needed, thus saving valuable production time. Since both the filler and pigment are stable products, these batches can be kept for long periods without any problems. However, it is best to prevent any moisture absorption from the atmosphere.

Bibliography

[1]   Defonseka, Chris. "Polymeric Composites with Rice Hulls." Pp. 93–100, De Gruyter 2019.
[2]   SpecialChem-"Pigments for Plastic Colorants." https://polymer–additives.specialchem.com/pigments-for-plastics.
[3]   Drumond, Farah Maria, Rodrigues de Olivera, Gisele, Ferraz, Elisa, Cardoso, Juliano, Zanoni, Maria, deOliveira, Palma. "Textile Dyes: Dyeing Process and Environmental Impact: Intech Open." January 16th 2013.
[4]   BASPAR Chemical Co. Iran-Artificial leather pigment pastes-www.basparlia.com/artificiallea therpigmentpastepvc.aspx.
[5]   SMOOTH-ON Inc. Colouring Urethanes and Silicones. www.smooth-on.com/support/faq/147.

Chapter 4
Selection of fabrics/base materials for coating

4.1 Introduction to textiles

Textiles, which are also called fabrics, are types of materials composed of natural or synthetic fibres or in some cases, combinations of them. Basic types of textiles include animal-based material such as wool or silk, plant-based materials such as linen and cotton and synthetic materials like polyester and rayon. Textiles are often associated with the production of clothing. In the manufacture of artificial leather, fabrics made of different fibres are used as backing for polymeric coatings. Basically, artificial leather is a combination of coatings and fabrics, with different surface finishes. Due to the importance of these fabrics which contributes a high percentage to the quality and properties of the final product, the author presents basic data regarding fabric constructions for the benefit of the readers.

The word 'textiles' was originally used to define a woven fabric produced by spinning raw fibres like wool, flax, cotton or others to produce long strands and forming them into a fabric by weaving, knitting, crocheting, knotting or pressing the fibres together (felt). Textiles fabrics are *planar structures* produced by interlocking yarns in some manner. These yarns in long strands are made up of individual long molecular chains of the discrete chemical structures. The arrangement and orientation of these molecules within the fibre, as well as the gross cross-section and shape of the fibre (morphology) will affect fibre properties. Usually, the polymeric molecular chains found in fibres have a definite chemical sequence which repeats itself along the length of the molecule.

For manufacturing artificial leather, different base materials (textiles) made from cotton, polyester, rayon, nylon and others are used, with cotton textiles probably being the most popular for direct polymeric coatings, while the others in the form of synthetic knitted textiles will give better results in the form of being lighter and more flexible when processed with coatings using the indirect or transfer coating systems.

Due to constant research and development, the world of textiles can now seriously consider two new plant-based fabrics, *bamboo fabrics* and *banana fabrics*, and welcome them into the family of textiles. While bamboo fabrics have made great strides as clothing and domestic applications, banana textiles are probably still being used for local craft. From the qualities of both products being marketed, it is apparent that they have much better potential for wider uses.

The difference between the woven and knitted fabrics is mainly in the weave, with the woven materials having a close weave (less flexible) and the knitted fabrics having more open weaves and very flexible. Both types, if coloured/dyed to match the colour of the coatings, will give the finished products a much better aesthetic look and feel. In general practice, woven fabrics are used for direct polymeric coatings,

https://doi.org/10.1515/9783110716542-004

while knitted fabrics are used for indirect/transfer coatings for foamed artificial leather. Knitted fabrics cannot be used for direct coatings due to their excessive open weaves which are too porous, while the woven fabrics although also porous to a much lesser degree can be used with a thickening agent for the base layer.

The basics of weaving involve the use of strands of yarn, one lengthwise (warp) and another across at right angles (weft), in a more or less up and under pattern to form a material. The *yarn count* (yarn thickness) also plays an important role in determining the weight of the final fabric. The fabrics will have *selvages* along the length on either side to hold the weaves or knits together. Unlike the materials made for the garment industries, for artificial leather manufacture, the length of each material/roll should be as long as possible, possibly 200–300 m, to enable easy continuous coating. If short pieces are joined to make a roll, wastages will occur at the joints.

Since most manufacturers will adhere to international standards, properties like weight per square metre, tear strength, width, flexibility and colour fastness for dyed cloths will be significant for all base materials used and an artificial leather producer will request these minimum properties from manufacturers/suppliers of base fabrics. These base materials, whatever the types used, are called *substrates* on which the polymeric coatings will be applied. To obtain smooth coatings, the base materials must be even, free of foreign matter, knots and of consistent thickness. Depending on the weave construction, a certain amount of shrinkage in width (weft) is to be expected on the coating machine due to the warpwise tensioning. Therefore, a coating processor will start with a wider width to arrive at the final desired width, making an allowance for side/edge trimmings also, if any.

Even for artificial leather, when a coated textile fabric is valued, comfort is considered as a fundamental property. Although, the type of coating used plays a major role in the final quality of the product, final quality also depends heavily on the fabric on which the coatings are built. It also depends on the chemical and physical properties of the fibres used to make yarn which is then woven or knitted into fabrics.

According to international standards, some of the basic specifications required when manufacturing artificial leather are – fabric weight, ageing, thickness, yarn count, shrinkage factor, tear strength and others, while from an end-user's needs properties such as air permeability, water absorption factor, flexibility, thermal resistance and others will be of value. Table 4.1 shows the basic constructional parameters of fabrics in general:

Table 4.1: Constructional parameters of fabrics.

Fabric type	Weave structure	Fibre type	No. of fibres/cross section	Warp density	Warp count	Weft density	Weft count

Here, it is worth mentioning again that whatever type of fabric is used, at least the coating surface should be smooth and free of blemishes like knots, fibres sticking out, uneven thickness across the weft (width) and warp (length), concentrations of dyes (globules) in dyed fabrics, frequent joints to make up the needed long lengths or any other. Generally, a coater would work closely with the fabric suppliers to manufacture the type of fabrics required.

4.2 Choice of substrates

Popular textile substrates used for polymeric coatings are: cotton, polyester, viscose, polypropylene (PP) and also blended fabrics made out of combinations of these fibres. Characteristics such as tear strength, tensile strength, dimensional stability, porosity and flexibility among others are important. Depending on the international standards being used by an artificial leather manufacturer, the standards for particular applications may differ and therein, a manufacturer may need base fabrics with particular specifications. In order to meet required specifications for the final product/products, it will not depend only on the base fabrics being used but a combination of both the fabric and the coatings as well. From a technical angle, one may say that the quality of the coating is the more important one. The strength and weight of a base fabric will depend on the construction method, the size and weight of yarn and the number of yarns per unit area of the fabric.

There are different polymeric coating systems with *direct* and *indirect* coating systems forming the base systems. Indirect coating systems are also known as *transfer coating* and a new system called *extrusion coating* has been developed by Davis Standard, which will be discussed under manufacturing methods in Chapter 8. In general, direct coating systems will use woven backing fabrics, while knitted fabrics will be used on indirect systems. Woven fabrics, irrespective of the types of yarns used will have certain porosity and the polymeric base coat will have to include a 'thickening' agent to prevent seeping through. Knitted fabrics, although their knits (loops) will have 'large openings', will be different as they will be 'laid down' and laminated on to the final foam coat.

With textiles and fabrics taking precedence for wearing apparel for the ever-growing population the world over, in some countries, manufacturers of artificial leather may experience difficulties in securing the exact types of fabrics needed and have to make do with whatever is available. Importing their requirements is one other option but may have to meet extra costs.

Polymeric coated fabrics combine and enhance the properties of a textile/fabric for a wide range of applications with the fabric component providing tensile strength, tear strength and elongation control and the coatings offering protection against environmental exposures and aesthetic values in speciality applications. Since natural leather is very expensive and its supply sources are limited, polymeric coated fabrics have established themselves well in all relevant end applications. The qualities of artificial leather have advanced so much that sometimes even experts may find it difficult to distinguish between the natural and artificial ones. Table 4.2 shows some of the fibres commonly used for fabrics for coating.

Table 4.2: Fibres for substrates for coating.

Fibre	Advantages	Disadvantages
Cotton	– No bonding agent – Low thermal shrinkage	*Low shrinkage per weight ratio *Absorbs moisture *Vulnerable to mildew, rotting
Polyester	– Low shrinkage – Relatively inexpensive	– Low moisture absorbency – Limited resiliency
Nylon	– Resistant to mildew, rotting, insects – Good elasticity and resilience – High abrasion resistance	– Low UV resistance unless protected – Fabrics may sag due to moisture absorption – Relatively expensive
Polyethylene, polypropylene	– Good thermal absorbency – Inexpensive – Chemically inert	– Low melting point–PE – Coating adhesion difficulty for some substances
Aramid	– Very high tensile strength – Good FR properties	– Degraded by UV or sunlight

Note: Some substrates or backing fabrics may be a combination of some of these fibres, while advanced technologies are now using plant fibres.

4.3 Woven fabrics

At the beginning when artificial leathers were produced, woven fabrics were commonly used and very popular with manufacturers. There are many different fabric constructions which are needed by the very wide range of end applications such as synthetic leather, clothing, fashion wear, industrial fabrics and transport coverings. However, only a relatively very small number of fabric constructions are employed for polymeric coatings. Examples are plain weave, twill and basket weave. The plain weave is by far the strongest because it has interlacing of the fibres used. Twill is also used for coatings with special effects.

Weaving is the oldest method of making yarn into different fabrics. While at the beginning, the weaving process was done by simple hand operated and then by semi-auto machines. With rapid advancement in weaving technology, modern day weaving processes are more complex, faster and automated. However, the basic principles of interlacing yarns remain unchanged.

On a loom, yarns threaded lengthwise are called the *warp* forming the base skeleton of a fabric. They usually require a higher degree of twist than the threading of the filling yarns interlaced at right angles, which is called the *weft*. Traditionally, cloths/fabrics were woven by a wooden shuttle that moved horizontally back and forth across the loom, interlacing the filling yarn with the horizontal lengthwise warp yarn. Modern mills use high-speed shuttle-less weaving machines that can produce a variety of different fabrics at very high speeds.

As an example, a rapier-type weaving machine has metal arms or rapiers that pick up the filling thread and carry it across the loom, where another rapier picks it up and pulls it across the rest of the way. Some other advanced weaving machines may use compressed air to interlace the weft yarn across the warp yarn. One of the advantages of using these machines, in addition to high speeds of production, is the relatively quiet operation.

There are three basic weaves with numerous variations and cotton can be used in all three. The plain weave (suitable for artificial leather coatings) consisting of the filling yarn (weft) is alternatively passed over one warp yarn and under the next is very popular for some basic clothing apparel. The twill weave in which the yarns are interlaced to form diagonal ridges across the fabric is used for sturdy fabrics like denim, ticking, gabardine and others. The satin weave, probably the least common of the three, produces a smooth fabric with high sheen. Used for cotton 'sateen', it is produced with a fewer yarn interlacing and with either the warp or the filling yarns 'dominating' the 'face' of the fabric. In advanced machine systems, optical scanners will continuously monitor, searching for flaws as the fabric emerges, with immediate print-outs showing the locations for remedial action during inspections for quality.

4.3.1 Micro-fibre polyester woven fabrics

One of the most important developments in the synthetic fibre industry is the production of extremely fine fibres called microfibres. In the production of artificial leather, basic important aspects are comfort and durability, whether it is used for furniture upholstery, footwear, automobile interiors or other. Towards this end, the two main contributors are the backing fabric (substrate) and the type of coating.

Micro-fibre fabrics are now being increasingly used in many different industries and particularly polyester micro-fibre fabrics in the artificial industry. Their popularity for end-uses throughout the world is due to their fineness, lightweight, durability, dependability, wrinkling resistance, breathability and their ability to be engineered to suit particular end-uses.

Tightly woven fabrics produced from micro-fibre yarns will have a very compact structure due to small pore dimensions between the fibres inside the yarns themselves. Polyester fabrics are slightly stiffer than cotton fabrics, but can be coated in such a way as to be very flexible and to drape well if coated fabrics are to be used for fashion wear. In Section 4.7, more information is provided about standard polyester fabrics.

4.4 Knitted fabrics

Knitted fabrics are used in the manufacture of foamed polymeric coatings using the indirect coating systems. Due to its highly open knit and porosity, these fabrics cannot be used for direct coating methods. Although cotton knitted fabrics can also be used, manufacturers of artificial leather prefer synthetic knitted fabrics due to their high flexibility, stretchability and elongation and light-weight properties. Artificial leathers made with knitted fabrics as the backing material are of a higher quality than the ones made with knitted or woven fabrics. Coated products with knitted synthetic fabrics will cost more but will fetch higher revenues as they will be used for high-end applications.

Knitted fabrics, unlike woven fabrics, are made by a 'knitting' process of two yarns forming loops in each course of the fabric knit. Knitting machines form loops of yarn with many pointed needles or shafts. The vertical rows of loops are called ribs or wales and the horizontal rows of loops are called courses. Knitting is the construction of elastic and porous fabrics by interlocking yarns by means of needles. Knitted fabrics can be made much faster than woven fabrics at probably comparatively less cost. These fabrics are generally light in weight, strong and their tendency to resist wrinkling is another plus as backing material for polymeric coatings. Since these fabrics are used as 'laid-downs' on the backs of the slowly moving foam coats, the high porosity is not a problem and is in reality an advantage. However, the lamination process must ensure that the fabric is not pressed down too much into the foam layer.

Because of the importance of knitted fabrics in the manufacture of high-end artificial leathers, the author presents below some of the variety of knit constructions, which gives a polymeric coater, a wide range of knits to choose from to suit different types of productions.

4.4.1 Knit schematics

Knitted fabrics are produced by two general methods – *warp knitting* and *weft knitting* with each method producing a variety of types of knitted fabrics. The knits of knitted fabrics are as follows:
– **Weft knits**
    Single knits: single jersey, Lacoste

Double knits: rib knit, purl knit, interlock knit, cable fabric, bird's eye, cardigans, Milano ribs, Pointelle

– **Specialized weft knits**
  – Intarsia
  – Jacquard jersey
  – Knitted terry
  – Knitted velour
  – Silver knit
  – Fleece
  – French terry
– **Warp knits**
  – Tricot
  – Raschel

4.4.2 Knitted fabrics types

(a) **Flat or jersey knit fabrics** have visible flat vertical lines on the front and dominant horizontal ribs on the back of the fabric. The flat or jersey knit stitch is used frequently. It is fast, inexpensive and can be varied to produce fancy patterned fabrics. A major disadvantage of these knits is their tendency to 'run' if a yarn is broken. The flat or jersey stitch can be varied by using different yarns or double-looped stitches of different lengths to make terry, velour and plush fabrics.

(b) **Purl knit fabrics** look the same on both sides of the fabric. Many attractive patterns and designs can be created with the purl stitch. Purl knit fabrics production is generally slow and is made by knitting yarn as alternate knit and purl stitch in one wale of the fabric. The fabric does not curl and lies flat being more stretchable in the length direction.

(c) **Rib stitch knits** have stitches drawn to both sides of the fabric, which produces fabrics that have excellent elasticity. Rib knit fabrics are made by knitting yarn as alternate knit stitch and purl stitch in one course of the fabric. These fabrics are reversible as they look identical on both sides of a fabric. They can be made with both flat and circular knitting machines.

(d) **Interlock stitch knit fabrics** are variations of rib stitch knits. The front and back of interlocks are the same. These fabrics are generally heavier and thicker than regular rib knit fabrics that do not curl or ravel at the edges.

(e) **Warp knitted fabrics** are made on special knitting machines. Unlike weft knits, they are knitted from multiple warp knit yarns, with yarns forming loops in adjacent wales. These fabrics are identified with a pick glass. The face or upper side of the fabric has slightly inclined vertical knit loops, whereas the rear side of the fabric has inclined horizontal floats. They do not ravel. Warp knit fabrics are constructed with yarn loops formed in a vertical or warp direction. All the yarns are placed for a width of a warp, parallel to each other in a similar manner of placement of yarns in weaving.

(f) **Tricot knit fabrics** are made almost exclusively from filament yarns because uniform dimensions and high-quality yarns are essential for use with high-speed tricot knitting machines. Fabrics made thus have plain or a simple geometric design. The front surface of the fabric has clearly defined vertical wales and the back surface has crosswise courses.

(g) **Raschel knit fabrics** are produced from spun or filament yarns of different weights and types. Most Raschel knits can be identified by their intricate designs, the open-space look of crochet or lace and an almost three-dimensional surface effect design.

(h) **Knitted velour fabric** are pile jersey fabrics having soft protruding fibres on the fabric surface. Like knit terry, they are also made of an additional set of yarns making pile loops on the fabric surface. However, in velour, these pile loops are sheared evenly and brushed. Since these materials are used with coatings for luxurious apparels, these fabrics will be dyed the colour of the coatings.

The above shows a few different types of knits being used to make knitted fabrics. To obtain maximum benefits of the final artificial leathers for different end applications, a choice of knits would be very useful, combined with the choice of yarn like cotton, rayon, polyester, nylon or combination of different yarns for coatings with polymeric foams using the indirect or transfer coating systems, manufacturers of artificial leathers will prefer synthetic knitted fabrics, most times.

4.5 Warp-knitted backing fabrics gains popularity

During the recent years, the versatility of synthetic leathers has made great strides as advanced technologies are applied in their manufactures. In the family of artificial leathers, the foamed types are the most popular ones with increasing end applications. Here, warp-knitted backing fabrics are increasingly being used by coaters and many more. The specific features of artificial leather are helping to find new and different applications. The availability and addition of newer additives to the polymeric coating mixtures to meet almost any end application, extends the versatility and popularity of artificial leather to great heights. The textile backing fabrics are mainly responsible for determining the stretch and weight per unit area and strength of these synthetic leathers.

Artificial leather, or synthetic leather as they are called, are made to international standards and the overall coated thicknesses for general purpose products may vary between 0.4 and 1.2 mm. However, for special applications, for example for footwear manufactures, greater thicknesses may be required. Whatever type of yarn or knit is used, one must remember that during the coating process, temperature for PVC is around 180–200 °C, while for polyurethanes, it is much less probably in the range of 60–70 °C. If different layers of polymeric coatings are used, the temperature structures will be different. Since heat is the final curing medium for all polymeric coatings, it is to be expected that the strength of the backing fabrics used will be affected and thus the overall strength of the final product. Therefore, when selecting backing fabrics, a manufacturer of artificial leather must select suitable ones.

Choosing which textile to use will largely depend on the final end use. For example, woven fabrics have dominated the clothing and furniture industries until recent times. Warp-knitted fabrics were considered too elastic for upholstery and too thick to produce fashion goods and used for making bags, luggage and shoes. For these products warp-knitted fabrics have an excellent tear resistance and a soft handle. However, current trends, as shown by a qualitative analysis carried out in China as the largest synthetic leather manufacturer, show a share of 70–80% of the total furniture production.

Warp knitted fabrics are generally made on high-speed tricot machines. One type of material is raised in the weft direction; it is soft and is made exclusively from textured and un-textured polyester. The coating is usually applied to the un-raised side to achieve a soft handle. The exception to this is the synthetic leather made used in furniture. In this case, the coating is applied to the raised surface to give it the necessary strength. When these fabrics are made for use as backing substrates, they will have low stitch numbers and the textured yarns are processed in a different position to the warp-knitted ones produced for sportswear.

4.6 Bamboo fabrics

The term 'bamboo fabric' refers to a wide variety of textiles like linen, bedsheets, pillow cases, comforters and many others made from yarns extracted from the bamboo plant. Different types of bamboo fabrics have been made over thousands of years but only recently have advanced processes been able to make fine textiles from these hardy and fast growing plants.

Bamboo fabrics come as a mixed bag. Depending on the process used, some productions of bamboo fabrics are environmentally sustainable, while some processes will use hazardous chemicals and maybe harmful to the processing workers. The majority of bamboo fabrics produced is *bamboo viscose,* which is cheap to produce, although it has environment downsides and could pose workplace hazards if proper precautions are not taken.

Viscose is a term used to refer to any type of fabric that is made using the viscose method, developed in the early twentieth century. It is comparable to rayon, which is a semi-synthetic fabric that was originally developed to imitate silk, which was expensive.

Production of viscose rayon involves the extraction of cellulose from wood pulp. The wood is broken down to small pieces and then exposed to chemical solvents to remove the cellulose. There are quite a few processes used to make viscose rayon and almost all processes include the use of harmful caustic soda.

Alternatively, bamboo cellulose can also be produced with a closed-loop production process. This process does not chemically alter the structure of solvents used, which can be re-used several times which significantly reduces the environmental impact. Bamboo fabrics of very high quality are made with production processes that do no extract cellulose. The bamboo is first crushed and then put into a bath of natural enzyme and then washed thoroughly and spun into yarn. This yarn will generally have a silk texture and fabrics made with this yarn are sometimes called – *bamboo linen.*

In general, these types of fabrics can be practically used for every application in which cotton is used. Some consumers may even prefer these fabrics cotton due to its beneficial attributes. For example, bamboo fabric is highly breathable and it is also more stretchable than cotton. They are easy to weave and fabrics with high thread counts can be produced resulting in textiles often thinner than their cotton counterparts, while remaining similar or greater in tensile strength.

The relative environmental sustainability of growing bamboo has stimulated the production of bamboo across the world and only recently the real value of the bamboo fibres has been realized. Producing bamboo is even popular in the USA and Europe, since it can be grown in a wide variety of climates and grows very fast. However, China is the largest producer of this type of crop, being an integral part of Chinese culture for centuries. Some other countries that export bamboo products are India, Pakistan and Indonesia.

Genuine bamboo fabrics made using mechanical methods (best qualities) may cost a little more than cotton products but is more affordable than forms of luxury cottons like Egyptian cotton, Pima cotton and Supima cotton.

There are three main types of bamboo fabrics. They are as follows:

- **Bamboo viscose**: these fabrics are nearly identical to other grades of viscose. One of the main reasons for using bamboo in the production of these fabrics is lower manufacturing costs.
- **Lyocell types**: these types of bamboo fabrics are similar to viscose but are made with a closed-loop production method. In addition, the chemical structure of the cellulose used to make these fabrics are not altered in the production process, which means that they retain many of the beneficial qualities that are also noted in mechanically produced bamboo fabrics.
- **Mechanically produced**: these types are made from fine bamboo fibre and can be considered to be true 'bamboo fabrics'. It could be relatively expensive and time-consuming to produce but they offer far greater benefits when compared to cotton in certain applications.

Bamboo plants are generally accepted as eco-friendly plants. For instance, bamboo is incredibly easy to grow, grows very fast and can grow in almost any area, especially areas that are not suitable for other crops. One of the main challenges for manufacturers of bamboo fabrics to be used as backing or coating substrates is producing long lengths in roll form.

4.7 Polyester fabrics

Polyester is a synthetic fabric that is usually derived from petroleum. These fabrics are very popular textiles and are used in hundreds of different consumer and industrial applications. Chemically, polyester is a polymer primarily composed of compounds within the ester functional group. Most synthetic polyester fibres are made from ethylene, a constituent of petroleum, while polyester fibres can be made from other sources such as plant-based. Polyester fabrics are also known as *polyethylene terephthalate*. While some polyesters are bio-degradable, most are not and are a contributing factor for environmental pollution.

The four different types of polyester fibres produced are:
- **Filament:** these fibres are continuous fibres and produce smooth and soft fabrics.
- **Staple:** polyester staples resemble the staples used to make cotton yarn and usually spun into a yarn-like material.
- **Tow:** polyester tow is like polyester filament but loosely arranged.
- **Fiberfill:** fiberfill consists of continuous polyester filaments but they are produced in such a way as to produce the most possible volumes.

When blended with cotton, polyesters improve the shrinkage, durability and wrinkling properties of polyester materials. These materials are highly resistant to environmental effects, which makes them ideal for long-term use and outdoor applications. Polyester fibre was first developed for mass consumption by DuPont Corporation, which also developed other synthetic fibres like nylon. Polyester materials can be used for coating polymeric mixtures in a wide range of end applications, including very thin polymeric transparent coatings for clothing. Printed and coated fabrics are used in the fashion industry.

4.8 Nylon fabrics

Nylon fabrics are also used in polymeric coatings for many different applications. While knitted nylon fabrics are used for polymeric coatings using the indirect or transfer method, woven nylon fabrics can be used for direct coatings.

Coated nylon fabrics play a wide role in industrial applications such as tarpaulins, flame-retardant fabrics for airbags, trunk-roll covers, motor hood insulation, safety belts, upholstery, seating materials, water and oil-repellent finishes and many others. Due to the high quality requirements of these products, a coater will, in all probability work in close consultation with a professional supplier of coating systems such as Stahl, who provides full service with performance coatings for fabrics.

4.9 Aramid fabrics

Aramid fibres are basically made from polyimide and are classified as specialized fibres. Plain woven para-aramid fabrics have excellent thermal characteristics. In addition to the good insulation properties, they can withstand a peak temperature of 500 °C and a continuous temperature of 350 °C. Added to these are specially valuable properties like resistance to cutting, abrasion, tearing, acids and excellent mechanical properties. Basic colour of aramid fabrics is yellow but they are available in different versions, such as aluminized and fleece. These fabrics are particularly suitable for manufacture of protective clothing and thermal insulation.

A standard version is 100% para-aramid Kevlar® fabrics but there is a cross-twill woven version made by two simple twillings oriented in different directions. Particularly resistant to high temperatures up to 450 °C, it is used for reinforcement of protective clothing. Some of the coated versions include silicone, aluminium with flame retardants.

For coating purposes, a good alternative to 100% aramid fabrics is a 70 Panaox®/30% Kevlar® twill. The fibre is an oxidized, thermally stable, polyacrylonitrile fibre. These fabrics are good for use against sparks and projected materials but its mechanical properties are inferior to a 100% para-aramid fabric.

4.10 Polyethylene fabrics

Olefin synthetic fibres are made from polyolefin derived from ethylene gas. Olefins are plastics that feel oily or waxy. They are polymers, compounds made from non-aromatic carbon and oxygen molecules of propylene or ethylene gas that are strung end to end. The two most important polyolefins are polyethylene (PE) and PP.

The term olefin can apply to fabrics made from either PE or PP with PPs dominating the market. They are preferred because they have a higher melting point and are generally stronger and lighter than PE. Some common applications of PE fibres are twine, rope, carpets, interior upholstery of automobiles, shopping bags and so on. PP fabrics are more often used in apparels, draperies, curtains and others.

Olefin fabrics made of PE do not deteriorate from chemicals, mildew, rot, sweat, sun or weather. They are lighter than polyester fibres, having a lower melting point but are strong and resist soiling and staining. Since, they are resistant to dying, colour where desired has to be added during manufacture of the fabrics.

PE fabrics are strong and do not stretch, making coated fabrics ideal for camping ware, deck chairs, tarps, tents and others. Products like carpets marked 'olefin' will usually contain both PE and PP fibres. High-density PE (HDPE) that resists cuts, ripping and tearing and is impervious to water and airborne contaminants is commonly made into protective clothing for industrial purposes, where there are biological or chemical hazards. Black PE fabrics are used as gardening mulch: blocks light from weeds, yet, allows moisture in between its weave. However, the ability of PE fabrics to resist deterioration by bacteria, heat, weather and other forms makes them non-biodegradable.

4.11 Polypropylene fabrics

PP fabrics are made from fibres derived from the thermos-plastic polymer PP, which belongs to the olefin group. They are non-polar and partially crystalline. This versatile polymer was discovered more or less by accident by scientists looking for gasoline from propylene. Originally, PP was marketed under the brand name *Moplen*. However, it is more common to find this material referred to as PP or 'polypro' for short.

As the use of PP became more popular in a number of consumer and industrial applications, it was discovered that this type of plastic also showed potential as a textile. PP fabrics are non-woven textiles, meaning they are made directly from a material without any need for spinning of yarn.

PP fabric is one of the lightest synthetic fibres in existence and is incredibly resistant to most acids and alkalis. In addition, the thermal conductivity of PP is lower than that of most synthetic fibres, which means that it is ideally suited for cold weather wear. In some cases, PP fabrics may also be used to make sportswear but now more advanced versions of PP have been found to be more suitable. While this fabric's moisture-transferring properties are highly desirable for sportswear, the inability to wash these materials with hot water may be a negative. In addition, susceptibility of these fabrics to UV action may also pose problems in certain applications.

PP is processed into fabrics all over the world, the biggest producers of finished polypro fabrics is China. Other countries where large quantities of PP are produced are Germany, Italy, France, Mexico, Belgium and the USA.

According to statistics, PP fabrics may be relatively expensive as compared to, for example, polyester fabrics. However, the increased costs really apply to grades of PP fabrics designed for apparel, while other general qualities which are available in different textures and colours will be cheaper.

4.12 Sheeting fabrics

Sheeting fabrics can be 100% cotton or a blend of polyester and cotton. Although, pure cotton is more absorbent and breathable, it wrinkles and has to be pre-shrunk. Adding polyester to cotton makes it more stable, eliminating shrinkage and wrinkles. Cotton sheeting is available, either bleached or unbleached and generally in widths between 36 and 118 in. These fabrics are relatively inexpensive and available in long rolls (without seams) and ideal for polymeric coatings using a direct coating system, although a certain amount of shrinkage (width) is to be expected due to the tension applied on a coating machine and also a base-coat 'thickener' will have to be used to prevent the polymeric mixture seeping through the woven material.

An artificial leather manufacturer will have the choice of quality sheeting needed which will largely depend on the end application. The thread count determines how soft and strong the fabric is, with generally, the higher the tread count, the finer the fabric. For breathable and lightweight fabrics, 200–500 thread counts should suffice. Most sheeting fabrics are finished by mercerizing and/or singeing, with the latter meaning the burning off of tiny fibres that would be sticking out on the fabric surface. The best quality sheeting fabrics will be both mercerized and singed. These fabrics can be easily coloured and for artificial leather manufacture, a coater would use different coloured fabrics to match the polymeric coatings which will improve the aesthetic value tremendously.

4.13 Other substrates for coating

There are other substrates that are also coated with polymeric mixtures. Although, strictly, they cannot be classified as 'artificial leather', they are linked through the coatings being polymeric mixtures and the same coating equipment being used. Moreover, these products are an essential part of daily life.

Some of the common substrates used for polymeric coatings for consumer and industrial purposes are: paper, synthetic sheeting, handlooms and materials from plant-fibre sources. While, polymeric coated materials for tarpaulins, tents, camping gear, protective wear, packaging, embossed shower curtains and others are popular, a good example of an 'artificial leather' made from plant fibres are the products made by a Sri Lankan company specializing in leather made from the Hana plant (agave plant).

4.14 Synthetic leather release paper

An important component in the manufacture of polymeric coated artificial leather using the indirect coating systems, mainly for foams, although, solid coatings on release paper with fabric 'laid down' laminated fabrics is also possible. These release papers are very wide and tough paper materials, and are able to stand high temperatures and are re-usable several times. There are different types of 'release coatings' on one side and probably the best ones are the silicone coated ones, which are most used by coaters. The silicone-coated surface will have a 'deep' emboss pattern, deep enough to transfer the emboss pattern to the first coating (solid), also known as the 'skin', on which the foam layers are built on. After the final coating and lamination of the backing fabric, the product is peeled off the release paper, the released surface with emboss will be the top surface of the artificial leather.

Cotek Papers UK-Products specializing in release materials offer the following range of release papers with a wide range of physical properties for a number of applications:
- Super-calendared kraft
- Sized/Prime kraft
- PE coated kraft
- Clay coated kraft
- HDPE film
- Polyester film
- Vegetable parchment
- Greaseproof paper

These release papers are also known as casting papers. Although they are re-usable for several runs, it depends on certain inherent manufactured qualities of the paper. Basically, they are graded on their weight per square metre. The range of widths will depend on the manufacturer or supplier and will generally be available in rolls of up to 2,000 m. Each roll will have only one emboss pattern so that a coater will have to purchase several rolls with selected embossed patterns as needed. In order to produce a plain surface, the release paper will have no emboss and all release papers are available in matt, semi-gloss or gloss finishes. Other important features when selecting these release papers are tear strength, degree of release ability, maximum temperature usage range as the basics. There are three main types of coating technologies for silicone release papers as follows:
- Thermally cured solvent-less silicones
- Thermally cured aqueous emulsion silicones
- Ultraviolet radiation cured silicones

Common problems in using these release papers are; scorching due to excess heat and re-use, tearing of paper due to improper alignment or handling on coating machine, small particles or globules of under cured polymeric mixture sticking on the release paper surface, thus, affecting the finished surface.

Most coating systems for producing artificial leather are wide and long to accommodate a coating head, laminating station and at least two curing ovens plus a finished product take-up station. A coater will select release papers with widths that can be accommodated on the machine with a few inches to spare on either side. Since release paper rolls are large and very heavy, a coating operation may need and mobile overhead system to lift and deliver/take-off these rolls to the machine, although a forklift may suffice for smaller rolls when used. Figures 4.1 and 4.2 show the basics of a released surface with emboss and a typical roll of release paper:

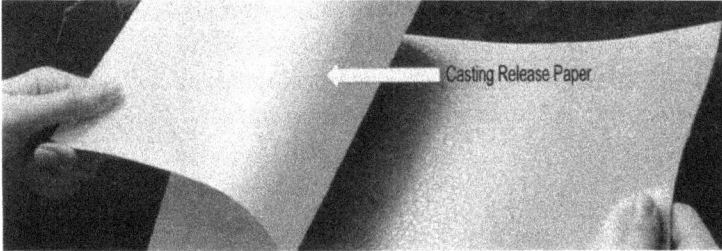

Figure 4.1: Embossed release surface. Photo courtesy of Taicang Andiya Technology Co. China.

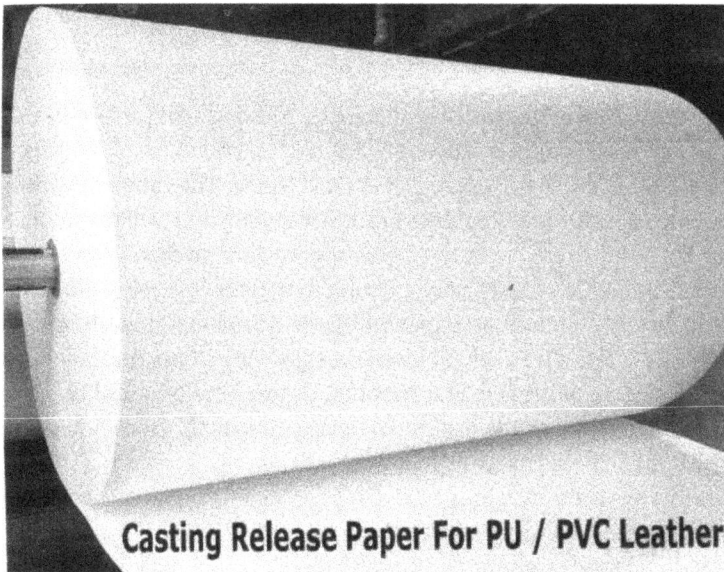

Figure 4.2: Large roll of release paper. Courtesy of Taicang Andiya Technology Co. China.

4.14.1 Example of general specifications for release paper rolls

Applications: PU/PVC artificial leather manufactures
Roll widths: 1,520–1,900 mm

Roll length: up to 2,000 m
Weight per square metre:
Material: virgin wood pulp
Feature: grease proof, easy release
Patterns/emboss: as per supplier range or per custom orders
Release finishes: matt, semi-gloss or gloss

4.14.2 Emboss patterns

A large manufacturer of release paper rolls may have up to 600 emboss patterns and only a few selected ones are shown below for the information of the reader:
- Crocodile pattern
- Suede pattern
- Calf skin pattern
- Shark skin pattern
- Snake skin pattern
- Lamb skin pattern
- Classic garment pattern
- Leopard skin pattern
- Pigskin pattern
- Plain surface pattern
- Random lines pattern
- Custom logo patterns

Generally, the animal skin patterns and high gloss plain patterns are used for handbags, luggage and shoes, whereas more subtle patterns will be used for upholstery. For clothing, plain and printed patterns may be preferred. Psychedelic patterns, which can be created by mixing of different colours of coating mixtures, can also be another option.

Since, these products can be classified as specialized products, the actual manufacturers may be limited but some of the sources of supply are The Wiggins Group Ltd. UK, Cotek Papers UK and Taichang Andiya Technology Co. China and so on.

Bibliography

[1] Hodakel, Boris. "What is polyester fabric: properties, how it is made and where".
September 28 2020, https://sewport.com/fabrics-directory/polyester-fabric.
[2] Hoyt, Richard. "What is polyethylene fabric?" https://www.hunker.com/13411444/whatispoly ethylene-fabric.

[3]   Hodakel, Boris. "What is polypropylene fabric: properties, how it is made and where."
      article October 03 2020, https://sewport.com/fabrics-directory/plypropylene-fabric.

[4]   TextileSchool. "Textiles – an introduction." article April 24 2020, https://www.textileschool.
      com/119/textile-anintroduction.

[5]   Kinge, A.P., Landage, S.M., Wasif, A.I. "Nonwoven for Artificial Leather." D.T.K.S. Society's.
      Textile &Engineering Institute, Ichalkaranji, India-ISSN 2278-6252 angSeptember 23 2020,
      https://sewport.com/fabrics-directory/bamboo-fabric.

[7]   Cotek Papers UK. Products: Silicone Release Materials, www.cotek.co.uk/products/products.
      htm.

[8]   "Cotton: From Field to Fabric." https://www.cotton.org/pubs/cottoncounts/fieldtofabric.
      cfm#.

[9]   Kettenwirk-Praxis: article published. "Warp-knitted backings turn synthetic leather business
      upside down," 6th September 2011, https://www.knittingindustry.com/warp-knitted-
      backings-turn-synthetic-leather-business-upside-down.

[10] IJESMR Journal. "Effect of Weave Structure on Comfort Properties of Microfiber Polyester
      Woven Fabrics."www.academia.edu/30537624, November 2016.

[11] Taichang Andiya Technology Co. China. www.alibaba.com/product-detailDY5298 Synthetic
      Leather Release Paper-Casting.

Chapter 5
Raw materials for polymeric coatings – PVC, polyurethanes and silicones

5.1 What are polymers?

Polymers are large molecules ('macromolecules') composed of repeating basic units, typically connected by covalent chemical bonds. The basic unit of a polymer is a *mer* and *poly* means many from the Greek word *polymeros*. Natural polymeric materials are: shellac, amber, natural rubber, cellulose – the main constituent of wood and paper. These materials have been used for centuries.

In due course, as technologies developed, the 'plastics' world saw the emergence of synthetic polymers mainly from crude oil at first and then from other sources. The word *plastics* is derived from the Greek word *plastikos* and can be described as a material that can be used in many ways by applying heat and pressure, generally for moulding purposes and the addition of plasticizers and additives to make plastisols of desired viscosities for coating purposes. The main starting source for synthetic monomers such as vinyl, styrene, ethylene, propylene and so on is liquefied ethylene gas as a by-product during the refining of crude oil. To form polymers, these basic monomers go through a process called *polymerisation* which can be done in different ways, the two basic processes being *addition* and *condensation* polymerization.

Polymers can be divided into two main categories – *thermoforming* meaning they can be re-used a number of times within certain limits and *thermosets* meaning they cannot be re-used once moulded or otherwise. While polyvinyl chloride (PVC) belongs to the thermoforming group, the polyurethanes (PUs) and silicones belong to the thermosetting group. There are three basic categories of polymers: *homopolymers, copolymers and terpolymers* as shown below:

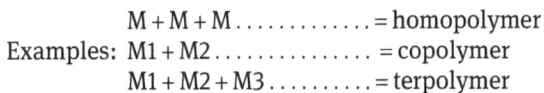

$$M + M + M \ldots\ldots\ldots\ldots = \text{homopolymer}$$
$$\text{Examples: } M1 + M2 \ldots\ldots\ldots\ldots = \text{copolymer}$$
$$M1 + M2 + M3 \ldots\ldots\ldots = \text{terpolymer}$$

M represents a basic monomer and when the same ones are combined the result will be a homopolymer, while, when two different monomers are combined will give a copolymer and three different monomers will produce a terpolymer. An 'elastomer' can be described as a polymer that can be stretched ≥200% of its original length.

https://doi.org/10.1515/9783110716542-005

5.2 What are polymeric coatings?

Polymeric coatings are coatings made with elastomers or polymeric materials to protect surfaces from corrosion, environmental hazards or other. These surfaces can be metal, glass, textiles, canvas, protective wear, insulation material, electrical cables and so on. There are many forms of coating methods depending on the products to be coated like dip coating, calendaring, spraying, direct coating, transfer coating and extrusion coating.

Speciality coated fabrics combine the beneficial properties of a textile and a polymer, synthetic or natural, the textile component providing tensile strength, tearing strength and elongation control and the coatings offering protection against the elements to which it will be exposed. Another aspect is comfort which is also very important when these coated fabrics are used as –upholstery and fashion wear. Application of these coated materials for handbags footwear and industrial applications like protective wear, transport covers and outdoor uses also makes these materials an essential part of daily life. Over the years, these coated materials have been known by many different names but they all can be classified under *artificial leather.*

In the case of artificial leather, polymeric coatings will take the form of *direct* or *indirect coating* on substrates like woven or knitted fabrics. Coatings for artificial leather are generally confined to PVC, PUs or silicone, which are the most popular. Depending on the end application and quality to be achieved, polymeric coatings can be of a single polymer or a combination. Colouring and surface finishes like matt, gloss, embossing and printing also plays an important role in the final quality and aesthetic values.

Examples of polymeric coatings include:
- PVC
- PUs
- Silicones
- Acrylic, epoxy
- Phenolic resins
- Nitrocellulose
- Polyethylenes
- Acrylonitrile–butadiene–styrene (ABS)

Polymeric coatings can be also applied to ceramics, wood, synthetic materials, both fabrics and films. They are generally, temperature resistant up to around 535 °F (280 °C) in most cases. Of the above, only the three most popular polymeric coatings are selected for presentation, which are PVC, PU and silicone.

5.3 Polymeric coatings for artificial leather

The range of popular polymeric coatings commonly in practice for various end applications is many as shown below. It is not possible to present all and in this chapter the author presents the chemical raw materials needed to manufacture PVC
coated, PU coated and silicone coated artificial leather, being the most popular
products due to their being the ideal substitutes for genuine natural leather, and
due to advanced technologies, very close comparisons can be achieved to natural
leather. The coatings can take the form of solid or foamed polymerics with a matt or
glossy surface finish. Embossed patterns, printing, lacquered or transparent surface
coatings will enhance the aesthetic values and also the versatility of a product. Naturally, each category will use different chemical components but some additives
may be common to all three. In addition to the individual component suppliers,
there are companies who will supply standard or custom-made polymeric coatings.
For economic reasons or convenience, a producer of artificial leather may opt for
the latter where purchases can be made on a just-in-time system, so as not to tie up
capital resources.

5.3.1 Polymers used for coatings

The following are some of the common polymers used for coating on fabrics by industry. They can be applied on fabrics (substrates) as solutions or dispersions in
water or as a solvent-based uncrossed mixture. If crosslinking of the polymer being
used is desired, it can be achieved after the disposition on the fabric, which will
improve the durability of the resultant coating to abrasion and its resistance to
water and solvent. The degree of crosslinking will be influenced by the nature of the
polymer itself, and the type and concentration used of the crosslinking agent:
– PVC
– PU
– Silicones
– Polyethylenes (PE)
– Polychloroprene (neoprene)
– ABS copolymers (nitrile rubber)
– Polytetrafluoroethylene
– Polyvinylidene chloride
– Styrene–butadiene rubber
– Isobutene–isoprene copolymers (butyl rubber)
– Chlorosulphonated polyethylene (Hypalon)
– Polyisoprene (natural rubber)

5.3.2 Acrylonitrile–butadiene–styrene

ABS is a terpolymer, made of three different monomers: acrylonitrile (A), butadiene (B) and styrene (S). By varying the percentage distribution of these monomers and adding different additives, it is possible to change the properties of ABS products to suit particular end-applications. Generally, the composition of the individual monomers are as follows:
- Acrylic nitrile 20–35%
- Butadiene 5–30%
- Styrene 40–60%

As antistatic agents, alkyl sulphonate and ethoxylate amine are added in concentrates of 1.5–3.0%. Fillers can be added in the range of 20–40%. Calcium carbonates can also be applied around 20–30%. If fire retardants are desired they can be added in small quantities. As lubricants, a choice between amide wax, zinc stearate and glycol mono-stearate. Antioxidants are normally added around 0.25–1.0%.

5.3.3 Rubber and thermoplastic elastomers

The classification –'rubber' includes a very wide range of materials. Some of the most common and important ones are as follows:
- Natural rubber
- Polystyrene-co-butadiene rubber
- Polybutadiene-co-acrylonitrile (nitrile rubber)
- Polyethylene-co-polypropylene-co-diene (EPDM)
- Polychloroprene (chloroprene rubber)
- Polydimethyl siloxane (silicone rubber)
- PU (PU rubber)
- Polybutadiene (butadiene rubber)

Rubber coated materials are also widely used in consumer and industrial uses. Some of these rubber types are also characterized as thermoplastic elastomers such as EPDM, silicone and PU which are polymeric materials used in the manufacture of artificial leather.

5.4 Chemical components for PVC coatings

The two main coating techniques for PVC artificial leather are *direct* and *indirect coating*. As the names imply, direct coating means prepared *plastisol mixtures* being directly applied to a substrate, for example a cotton fabric. Since these woven fabrics will have an open/porous weave, the first (base) coat needs a *thickening agent* to prevent the plastisol seeping through. The second coat may be a solid coat or one containing a blowing agent to form a foam layer. The final top coat (skin) will be a solid coat.

Indirect coating or transfer coating involves the coating of the first layer (skin) being coated onto an embossed release paper, followed by one or two coats of foam and then a knitted fabric being laminated on to the wet foam. The following are the basic chemicals needed for both types of coating manufactures.

– Polymers, co-polymers, graft polymers (filler polymers)
– Plasticizers
– Fillers
– Heat stabilizers
– Lubricants
– Thickening agents
– Blowing agents
– Rice hulls flour/rice hulls ash (thickening agents)
– Egg shell powder (filler)
– Embossed release paper
– Dyes and pigments
– Fire retardants
– UV (ultraviolet) agents
– Anti-microbial agents
– Crosslinking agents
– Matt or glossy surface agents

5.4.1 Copolymers and graft polymers

A copolymer, as explained above, will consist of two different monomers. Different grades of polymers can be used with the graft polymers functioning as 'filler' polymers. They can be varied in different proportions in a formulation but with the total polymer content based as 100 parts by weight on which the other components of a formulation are built to make suitable plastisols to suit each manufacture. The ideal would be to have an in-house laboratory and the services of a chemist or plastics technologist where an artificial leather producer could formulate different coatings to suit end applications. This type of infrastructure would suit large volume and sophisticated operational units, while the smaller producers could purchase their coating requirements from many suppliers who would also offer laboratory facilities for testing and short production sampling.

5.4.2 Plasticizers

Plasticizers are basically non-volatile organic substances (mainly liquids). When introduced into a polymer or an elastomer, it will improve the following properties:
- Flexibility
- Extensibility
- Processability

Plasticizers can be used as a single component or as combination with another plasticizer and will soften and improve the thermos-plasticity and the flow of a polymer by reducing the viscosity of the polymer melt or mix, the glass transition temperature (T_g), the melting temperature (T_m) and the elastic modulus without altering the fundamental chemical character of the plasticized material. For PVC, plasticizers are an essential component as the PVC monomer is basically hard.

Plasticizers are among the most widely additive in the plastics industry. They are also usually cheaper than other additives used for polymer processing. Plasticizers are most often used for PVC, the third largest polymer after polyethylene and polypropylene. However, significant amounts of plasticizers are also used in polymers like acrylics, polyolefins, PUs and others.

From the wide range of plasticizers available for the preparation of plastisols for polymeric coatings for artificial leather with PVC coatings, common plasticizers like DOP, DIOP, DBP can be used among others.

5.4.3 Fillers

Common fillers used in industry are mineral fillers and their main functions are to reduce costs. While the use of fillers can be an advantage in polymeric coatings and also enhance significant properties, fillers in moulded plastics may sometimes weaken the strength of a moulded part. For all applications, the important basics are particle size, moisture content and the threshold limit of use regarding a particular process/product.

Minerals commonly used as fillers in plastics are *calcium carbonate, talc, silica, wollastonite, clay, calcium sulphate fibres, mica, glass beads and alumina trihydrate.* For PVC polymeric coatings for artificial leather production, the author would recommend calcium carbonate.

Alumina trihydrate gives flame retardancy because it decomposes endothermically into alumina and water when heated above 220 C. However, its low decomposition temperature limits its use in resins like polyethylene and polypropylene which have relatively low processing temperatures.

Clay fillers improve electrical properties and processability of most polymer resins. Mica will boost a resin's mechanical properties but its use is limited due to its brown colour. Calcium sulphate and wollastonite are fibrous or needle-like fillers and can either slightly improve or degrade mechanical properties. Glass beads, silica, calcium carbonate and talc usually weaken plastics.

5.4.4 Heat stabilizers

PVC is one of the most important commercial polymers and its compounds have a great diversity of applications but since it is thermally unstable, certain additives are needed during processing. Thus, heat stabilizers play a key role in processing. The amount and type of energy input will vary considerably among the many different production methods and end-use applications. Mild degradation of the monomers can start even during polymerization and continue into the finished PVC raw material in storage.

5.4.5 Lubricants

These additives are used in polymeric coatings primarily for achieving fluidity among the different components used in a mix to achieve a well-mixed homogenous polymeric mix and also to enable smooth/ease of flow during the coating operation. Lubricants can be incorporated in a formulation as a single additive or more than one as a system.

5.4.6 UV stabilizers

Long exposure to sunlight (UV rays) can be harmful to any polymeric coating. PVC and PUs are especially prone to degradation and these coatings have to be protected. A sure early sign of degradation is change in colour, which could also happen in patches. In thicker coated leather, degradation can even cause surface cracking.

Whatever type of artificial leather is manufactured most will need protection from possible fire hazards. It may not be possible to completely stop the destruction of coated fabrics in a case of exposure to fire but most specifications will require that affected materials will not promote a fire. There are, of course, certain specially made coated materials made for fire-proof wear.

5.4.7 Dyes and pigments

They are generally used in polymeric mixes to enhance its aesthetic values. Most dyes and pigments are compatible with PVC and PUs and achieving good even colouring needs some effort. Any unmixed colour in the form of 'globules' will result in 'splay' during the coating operation or may even show up as a 'lump'. One technique to avoid this is to add the weighed colour content to the filler and make into a fine particle paste on a triple-roll-mill.

5.4.8 Blowing agents

They are used in both types of coatings *direct* and *indirect*. These additives are incorporated into the middle coats, where the polymeric mix is 'blown' under controlled heat conditions to obtain a cellular foam structure. To achieve high-quality uniform foam structures, it needs experience on the part of the coaters. Two of the main problem areas are in direct coating during the embossing operation care must be taken not to 'crush' the foam surface by too much pressure applied by the embossing roller. In indirect coating, when the knitted fabric is laminated on to the back of the foam surface, too much pressure will again result in 'crushing' the foam surface. One solution in both cases is to 'blow' the foam layer to a slightly higher level to compensate for the pressures that will be applied.

5.4.9 Thickening agents

When woven fabrics are used in direct coating methods, when the base coat is applied on the fabric the polymeric mix will seep through the open weave. To prevent this, coaters will add a thickening agent to the base mix. In indirect coating, thickening agents are not required as the knitted fabric should 'penetrate' the foam coat during lamination. Here, the degree of penetration is important to prevent easy 'peel-off' or separation.

5.4.10 Additives for matt or glossy surfaces

Artificial leather manufacturers may use additives to obtain a matt or glossy surface. In general, most polymeric surfaces give a fairly glossy or matt surface. If these finishes are important, the transparent coatings applied as a final surface protective layer should be highly transparent to achieve full effects.

5.5 Chemicals for polyurethanes

The following are a list of the basic chemical components that can be used to formulate suitable PU coating mixtures for making PU coated fabrics. The technology here is more complicated than for PVC, and if an experienced chemist is not available on a production floor, a coater may want to purchase suitable ready-mixed mixtures from many professional suppliers who can supply coating materials to any specifications. However, the question of shelf-life and costs will be factors to be considered.

List of chemical components:

- Polyols
- Graft polyols
- Isocyanates
- Bio-polyols
- Catalysts
- Blowing agents
- Surfactants
- Methylene chloride
- Pigments
- Fillers
- Anti-static agents
- Lubricants
- Heat stabilizers
- Plasticizers
- Cell openers
- UV stabilizers
- Others (for special properties)

Any of these can be combined, with the polyols and isocyanates as basics. It should be noted that grades used for PU flexible foams may differ for polyols for coating purposes. If details of these components are desired, a reader may refer to Smithers Rapra publication *Practical Guide to Flexible Polyurethane Foams* (ISBN: 978-1-84735-975-9 (e-book), 2013).

5.6 Silicone coatings

Silicone, also called polysiloxane, is of a diverse class of fluids, resins or elastomers based on polymerized siloxanes, substances whose molecules consist of chains made of alternating silicon and oxygen atoms. Their chemical inertness, resistance to water and oxidation along with stability at both high and low temperatures have led to a wide range of commercial applications from lubricating greases to electrical-cable insulation, bio-medical implants. Latest research and developments have allowed this versatile polymer to be used for coatings, especially for artificial leather where its softness, toughness, flexibility and elasticity permit the manufacture a high-end artificial leather, especially for the automobile industry.

For many years past, artificial leather has been defined as either PVC or PU coated fabrics. Both have their own strengths and weaknesses, but have served their designated end applications very well. In fact, they are still in great demand as imitation leather for natural leather. Their qualities have improved tremendously and since natural leather is very expensive, PVC and PU leather hold their own on being cost-effective also.

While PVC leather has been considered the 'workhorse', being even cheaper than PU leather, good quality and availability have contributed greatly in their being used for public place and transport applications, along with standard furniture, upholstery, medical and other applications.

The silicones differ from most industrial polymers in that the chains of linked atoms that make up the backbones of their molecules do not contain carbon, the characteristic element of organic compounds. This lack of carbon in the polymer backbone makes polysiloxanes into unusual 'inorganic' polymers, though in most cases two organic groups, usually vinyl, methyl or phenyl are attached to each silicon atom. A general formula for silicones is $-(R_2 Si O)_x$, where R can be any one of the organic groups.

The most common silicone compound is poly-dimethylsiloxane which can illustrate the central characteristics of their class. The starting material is metallic silicone, which is obtained from silica sand. Silicon is reacted with methyl chloride using a copper catalyst, and by reacting this compound with water, the chlorine atoms are replaced by hydroxyl groups. The resulting unstable compound – silanol – is then polymerized in a condensation reaction, the single-unit molecules linking together to form poly-dimethylsiloxane.

Polysiloxanes are manufactured as fluids, resins or elastomers, depending on the molecular weight of the polymers and the degree to which the polymer chains are interlinked. Some of their applications are as lubricants, hydraulic fluids and as emulsions for imparting water repellency to textiles, paper and other materials. Latest technologies have emerged as ideal coatings for artificial leather, which may be considered as superior to PVC and PU coated leather for some applications, even though they are more expensive.

Bibliography

[1] Galata Chemicals. www.galatachemicals.com/products.
[2] Special Chem. https://polymer-additives.specialchem.com.
[3] Machine Design. www.machinedesign.com/materials.
[4] "Practical Guide to Flexible Polyurethane Foams." Smithers Rapra publication, 2013.

Chapter 6
Concepts, theory of formulating and formulations

6.1 Concept

The basic concept of manufacturing artificial leather is the coating of a synthetic polymeric coating on a substrate, generally, on a fabric or textile to achieve a close imitation to natural leather. The fabrics or textiles can be made from natural fibres, synthetic fibres or bio-fibres from plants. The two most commonly used methods of coating are *direct* and *indirect* coating systems and now a new coating system called *extrusion coating*, as offered by Davis Standard, is available for this industry.

Whatever coating method is used, the principle concept remains the same, and for high-quality foamed polymeric coated artificial leather, an indirect/transfer coating system where knitted fabrics are used as the substrate, unlike woven fabrics that are used for direct coating systems. For high-quality artificial leather products, both foamed polyvinyl chloride (PVC) and polyurethanes (PUs) are suitable, while the latest silicone coated fabrics are proving to be a class of its own.

When we think of artificial leather, perhaps our mind may give priority for these products as for end applications like upholstery and furniture. The range of artificial leathers has a much bigger spectrum in that there are other equally big fields of applications like footwear, fashion wear, safety wear, tarpaulins, hand bags and raincoats. Although the basic principles of manufacture, for example coatings on fabrics/substrates, remains the same, production techniques will differ. Add to this, the coatings on other substrates like paper, synthetic films and materials from bio-fibres. Each will have its own technologies and coating techniques.

The concept and production of synthetic coated fabrics to manufacture artificial leathers requires a good overall knowledge of the polymeric mixtures being used, the coating techniques, the capabilities and limitations of the equipment being used and achieving the finishes desired, on which depends the quality and aesthetic values of the products. Equally important is a good understanding of the specifications to which these products are being made, which can be a two-tiered one-customer specs and international specs.

6.1.1 An overview of artificial leather production

The term 'artificial leather' came into being with the need for a substitute for natural leather, which was both expensive and dwindling in supplies. Since natural leather was processed from animal hides, protests from various groups also contributed largely for the birth of artificial leather which was made from synthetic polymers coated on cotton fabrics and later on other substrates also.

https://doi.org/10.1515/9783110716542-006

The fabrics also called textiles can be based on a number of different fibres, for example natural fibres like cotton or acetate fibres or synthetic fibres like polyamide, polypropylene (PP) and polyester or others. The fabrics are made by processes of weaving, knitting or as non-woven materials. These will form the backing fabrics in the manufacture of artificial leather and should be made available to the coater in long lengths and in roll form, preferably without any joints and surface blemishes like knots, fibre sticking out and so on. In most cases, a coater will request coloured (dyed) cloths to match the colour of the coatings from the suppliers, which will enhance the aesthetic value of the finished products.

These fabrics can be coated with polymeric mixtures such as PVC, PU, silicone, ethylene–vinyl acetate copolymers, butadiene copolymers, polyamide, acrylonitrile–butadiene–styrene (ABS) and thermoplastic olefins to give different types of artificial leather. These products for looks and touch will be ideal substitutes for natural leather of different qualities. With advances in technology and highly developed additives, manufacturers of artificial leather are able to produce very good imitations of natural leather and in most cases even an expert may find it difficult to tell the difference. Figure 6.1 shows embossed artificial leather.

Figure 6.1: Embossed artificial leather.

6.1.1.1 Artificial leather with PVC

PVC coated artificial leather is still probably the most common type of artificial leather in the market, although, other types like PU coated and silicone coated fabrics are considered as higher-end products. There are still other types of polymeric coated fabrics for speciality applications which can be classified under the umbrella of artificial leather and are not discussed in this presentation due to the many types involved.

PVC is basically a stiff homopolymer and in practice it will have to be modified to achieve the desired properties for products and in this case, for polymeric coating mixtures. General basic additives added for PVC polymers are plasticizers, stabilizers, fillers, lubricants and other additives to achieve a desired result. Plasticizers soften the PVC polymers from a stiff to a fluid state and added as a percentage of the PVC polymer taken as 100%. Softened PVC can form very large variations of different compounds or mixtures with other compatible polymers like ABS, synthetic rubber and acrylics.

The plasticizer, the key ingredient in a PVC mix, can be a single or a combination of plasticizers and can be as high as 70%. The most commonly used plasticizers are *phthalates* such as DIOP–di-octyl phathalate and others. With current environmental concerns, one may want consider other alternatives to phthalates such as adipates, citrates, phosphates, epoxidized soya bean oil and others but care must be taken to ensure whether the desired properties can be achieved with these alternatives over the tried and tested traditional ones.

Since the thermal stability of PVC is limited, a heat stabilizer or a stabilizer system has to be used. For many years, the use of heavy metal compounds were the norm but due to health and environmental concerns, especially the lead-based compounds, organic and inorganic metal compounds like calcium-, barium- and zinc-based have replaced them as primary ones with still others as secondary ones.

The biggest challenges artificial leather has to face other than flexibility, durability and feel are the effects of light. For starters, excess exposure to heat will make the plasticizer content from the surface of the polymeric coating to evaporate gradually and make the coating brittle and liable to cracking. A good example is the interior of automobiles, where a good percentage is made up of artificial leather. This can lead to decomposition of the PVC also. To prevent this, a polymeric coater will include an effective combination of an ultraviolet (UV) stabilizer/antioxidant system in a polymeric formulation.

There are four basic types of antioxidants which are commonly used for many different types of plastics which can be used ranging from around 0.5% to 20%. They are as follows:
- Phenols, monophenols, bisphenols, thiobisphenols and polyphenols constitute the largest groups of primary antioxidants. In polyphenols like polyethylene (PE) and PP, they are typically used in amounts of 0.05–0.2 %, while in styrene-based materials, they are applied up to 2.0%.

- Amines and amine-based antioxidants are especially used in synthetic rubbers in amounts between 0.5% and 3.0%.
- Phosphites are used alone or together with phenols or amines. In polyolefins, amounts of 0.05–2.0% are used.
- Thioesters are important to prevent changes in molecule mass at long-term exposure to high temperatures.

Other additives are barium, calcium and zinc compounds (0.05–3.0%), pigments/colouring agents – up to 10 parts per 100 parts of PVC – fillers (20–60 parts PVC), flame retardants up to 10% and lubricants and biocides.

General components for a standard polymeric coating would comprise a co-polymer, graft polymer, plasticizers, heat stabilizer, lubricant, filler, thickening agent for base coat, colour and additives to obtain special properties. These would produce a solid coating and if a foamed coating is desired, a compatible blowing agent can be included. The thickening agent is only for the base coat as most woven fabrics would have a fairly 'open weave' and the polymeric mix would seep through.

6.1.1.1.1 An overview of general compounding chemicals used with PVC resin

The essential ingredients in a PVC coating formulation are:
- PVC resin
    - Suspension grade
    - Paste grade
    - Copolymers
- Primary plasticizer
- Secondary plasticizer
- Stabilizers
    - Heat stabilizers
    - Light stabilizers
- Lubricants
- Fillers
- Pigments
- Additives (special properties)

There are four basic types of PVC resins as follows:
1. Suspension grade
2. Emulsion grade
3. Bulk polymerized
4. Copolymers

6.1.1.1.2 Suspension-grade PVC

Suspension grades of PVC are made by polymerising droplets of vinyl chloride monomer suspended in water. When polymerization is complete, the slurry is centrifuged and the PVC 'cake' is gently dried by special heating systems so as not to expose the unstabilized resin to degradation. According to general practice, the particle size of this resin will range 50–250 microns and will have porous structures which will readily absorb plasticizers. The structure of the PVC particles can be modified by selecting suitable suspending agents and polymerising catalysts.

6.1.1.1.3 Emulsion grades

Emulsion polymerized PVC is what paste-grade resins are and are widely used for plastisols. Paste-grade resins have very fine particles and are produced by spray drying an emulsion of PVC in water, very much like how milk powder is produced. Paste-grade PVCs require much more energy to produce and will cost more than suspension grades.

Paste-grade resins carries the emulsifying chemicals and catalysts with it. It is, therefore, less pure than suspension-polymerized PVC. The electrical properties of paste-grade resin plastisols are poorer than that of suspension grades. Paste-grade resins are compact structures and will not absorb much plasticizer at room temperatures. Temperatures in excess of 160–180 °C are needed to make the resins absorb plasticizers during curing. Paste-grade resins are widely used for cushion vinyl floorings of wide widths. Different layers of specially formulated pastes are coated on a suitable substrate (direct coating) or on a release paper (indirect/transfer coating). The layers are continuously fused in long heating ovens and rolled up after separation from the release paper. The rolled flooring material can have a tough transparent protective wear layer over printed/embossed and foamed layers which will have thick highly filled base layers to build up the required thickness levels. Many types and qualities of high-end vinyl floorings can be made in this fashion.

6.1.1.1.4 Bulk polymerized PVC

Bulk polymerization will give the purest form of PVC resin as no suspending or emulsifying agents are used. These PVC grades are mainly used in transparent applications. There are mainly made in the lower K-value (mol. wt./polymerization degree) groups and ideal for PVC foils, blister packaging material or calendared/extruded transparent films which are best produced from lower K-value groups. According to reports, advances in suspension resins technology has edged out bulk PVC in the recent past.

6.1.1.1.5 Copolymer PVC

Vinyl chloride is copolymerized with monomers like vinyl acetate to give a range of resins with unique properties.

The range of coating thicknesses can be from 0.4 to 1.2 mm or thicker, depending on the end application. In the process of solid polymeric coatings, using a direct coating system, the final thickness of the product is achieved by the thickness of the fabric and build-up of two or three coats, with the base coat being the first and the last coat being the top coat or 'skin'. In some cases, a coating allowance may have to be made to compensate possible loss of 'height' during embossing. These surfaces are generally embossed with different patterns or printed. To improve the quality of the artificial leather, a coater will use an additional transparent or lacquer coat on the surface which will improve heat and light stability, also preventing the evaporation of the plasticizers due to prolong exposure to heat. A good example is the interior of an automobile. The applied lacquer can take the form of solutions of polyacrylate and PVC in organic solvents in a layer thickness of 3–20 μm. The standard PVC mix without the filler and colour will also give a smooth transparent coating.

6.1.1.2 Artificial leather with polyurethane

PU coated fabrics as artificial leather is generally considered as better quality products as compared to PVC coated products due to their flexibility, softness and being a very close in quality and feel to natural leather, although more expensive.

In the production of PUs, both aromatic and aliphatic isocyanates are used and the finishing quality of a product is to a large extent determined by the grades of polyols used, either singly or in a combination. Polyether and polyester polyols are used most often but some productions may use polycarbonates.

For flexible PUs, 63% polyol, 34% isocyanate and 3% additives can be used as a general rule but for artificial leather productions different parameters may apply. A PU formula for artificial leather will probably include a pigment/dye, blowing agent (foam), a thickening agent (direct coating base) and others as desired. There is a large selection of polyols and isocyanates available which enables a coater a wide range of properties of finished products possible. In practice, toluene diisocyanate and diphenylmethane diisocyanate are used most often with the latter gaining ground because of a lesser toxicity. If the polymeric material is to be extra soft, plasticizers like phthalates can be also incorporated.

PU coated leather is made by both direct and indirect coating methods to produce solid coated fabrics and also foamed coated fabrics and embossed, printed or lacquered as desired. PU coats do not need any softeners as a component in a formula but because of its 'soft' nature, care is needed at the time of embossing where a patterned roller is gently pressed on a warm/hot final coat surface which will compress the surface. A coater may want to have a slightly thicker overall coating to compensate for this loss in height.

The preferred method for top-quality PU coated artificial leather is the indirect/transfer coating method, where the coatings are done on a 'reverse' basis unlike direct coating. The top coat of the 'skin' is first coated on to an embossed release paper and then the required thickness is built up in layers (two or three coats) which will have a blowing agent which will decompose as it goes through the hot ovens and form a cellular foam coating. A knitted fabric is laid down and laminated on to the last coating and here again, the compression must be minimal to prevent loss of overall thickness.

6.1.1.2.1 Aqueous two-component polyurethane coatings

Since PU coated fabrics came on the market there has been many varieties. PU-based coatings have an established place in the coatings industry and in some applications they dominate the market. Development of a new generation of water dispersible polyisocyanates as well as improvements in application technology have contributed greatly to this. For industrial finishing and automotive refinish, there are grades with special properties such as rapid drying coatings with well-balanced range of, properties at low volatile organic compounds (VOC) levels.

There are a number of options for producing low VOC coatings. Waterborne PU-coatings have already taken a significant share of the market and has so much potential for further development with a bright future. Modern waterborne products must at least match property profiles of solvent borne systems. Until recently, it was not possible to produce two-component PU coatings. This was mainly due to the obstacle of the undesired secondary reaction of the water with the hardener. As with most polymeric coatings as developments increased and as new products arrived on the market, we have to accept that new challenges have to be faced with regard to the needs of suitable coating systems, plants and manufacturing technologies.

Whereas in the past, maximum performance and appearances were the main market requirements, resulting in the developments and increase of solvent borne PU systems usage, due to environmental concerns, the current demand is for coatings with reduced harmful emissions. Since PU coatings produce high-quality products for which there is an increasing demand, the manufacturers of PU raw materials and the coating industry as such to include manufacturers of coating systems have been forced to increase their efforts to develop suitable products and technologies to meet these new demands.

PU coatings yield a high level of quality and with PU coated artificial leather, a very close resemblance to natural leather can be achieved. They combine some important properties with resistance to solvents, chemicals and good weather stability. New technologies permit formulating both clear coats and because of their good pigment wetting properties, pigmented topcoats which can yield high gloss, high-bodied films with excellent flow properties. These films will have outstanding mechanical properties and will provide an ideal balance of hardness and flexibility, even at low temperatures.

6.1.1.3 Silicone coated artificial leather

For many years, artificial leather has been defined as either PVC or PU coated fabrics. As we know, they are two different polymer-based types having their own strengths and weaknesses. Both are used widely in many applications as a great substitute for natural leather which is expensive and anyway in short supply. PVC leather, being the first, was considered the go-to 'workhorse' suitable for almost any application.

With the advent of PU coated fabrics, they were immediately accepted as and recognized as a higher quality product due to their softness, unique surface feeling and versatile design options, although more expensive. However, both types continue to be in great demand and today, with advanced technologies, the qualities have improved tremendously, so much so, that one might say that natural leather made from animal hides are not necessary.

To strengthen this claim, the arrival of silicone coated artificial leather for the leather industry from 100% silicone coated to hybrids to specialized topcoats on fabrics have given the artificial leather industry a tremendous product. According to many, these fabrics/textiles are arguably the most exciting polymeric coated products in the market today, with many possibilities, to join the family of coated fabrics/textiles, enhancing an already established market. As with any new product platform, producers of these materials are making available access to end-users about the special features and benefits of their products.

Silicone is basically derived from sand which is processed into organic silicone monomer and silicone polymer. The silicone monomer and the silicone polymer are compounded and then mixed and goes through a casting system through coating and heating stations. The casting system is very efficient without off-gassing and thus, no pollution. There is no odour since silicone is odourless and silicone coatings probably offer the least environmental hazards among the range of coated fabrics, although silicones are not bio-degradable.

Silicones are 'rubber' type of materials with unusual molecular structure with changing silicon and oxygen atoms. Each silicon atom carries one or several organic groups, normally, methyl or phenyl groups. The most used silicone is polymethyl siloxane, which due to its eminent bio-compatibility is one of the most used, especially, for medical purposes such as implants.

Sileather is a popular silicone coated artificial leather, similar to PVC (vinyls) and PU coated fabrics. Manufacturers of this brand claim their superiority as an environmentally friendly alternative to traditional coated fabrics with a long list of benefits and advantages over these fabrics. This is based on their developing and refining their own silicone raw materials for many years. According to them, silicone is in the middle of being an organic and inorganic material and can be categorized as an *elemental polymer* with a main molecular chain and mainly composed of a Si–O–inorganic base unit with its side chain connected through the silicon atoms and other organic groups. Therefore, silicone will have both the stability of an inorganic material with the elasticity and plasticity of organic materials.

Perhaps one of the most prominent performance of silicone materials is the high safety advantage as the material is widely used for baby pacifiers. Another performance characteristic is its ability to endure a wide range of weather conditions. Silicones also excel in performance as electrical insulation, stain resistance, flame resistance, waterproofing, durability, chemical resistance, eco-friendly and so on.

According to Sileather, they use some of these special features of silicone in their coated fabrics which can be used for applications both indoor and outdoor. Their silicone Sileather is made, according to them, with a proprietary 100% silicone recipe, specially formulated. These fabrics are made for outstanding performances such as scratch resistance, sagging resistance, UV resistance, chemical resistance, easy to clean, flame resistance and very soft textures and these characteristics are achieved inherently, without the use of any added chemicals.

In the polymerization process catalysts such as platinum or peroxides can be used. This versatile material as a filler, for example, silicates, quartz powder, talcum powder and calcium silicates, up to about 38% (w/w) can be used.

Silicones are stain-resistant and coatings are easily cleaned when necessary with simple soap and water or light chemical solutions but it is advisable to consult producers if the latter are used. One drawback with silicone coated fabrics is that these surfaces cannot be embossed, so most embosses come with embossed release paper, printing or digital printing.

Are silicone coated textiles more expensive than others? –Yes, the raw materials are expensive and the process is expensive as they are all organic in nature. The release papers which provide the desired grains/emboss cannot be used as often as for other polymeric coated fabrics and must be replaced more frequently. The estimated production speed is about 40% slower than for other conventional coating processes. However, as production technology improves, it is expected that the prices will come down.

6.1.1.4 New advanced polymer systems

The polymer industry is rapidly developing through constant research and development to meet the frequent demands of the various sectors of end applications. This is also very true for artificial leather end users. Over the years, especially due to the demands of animal rights groups and escalating prices, the manufacture of natural leather has diminished to a great extent and artificial leather has been filling these needs with rapidly increasing volumes of productions to meet all needs of the various sectors.

However, due to the needs of ever-increasing challenges being posed by sectors such as the automobile, furniture upholstery, space travel, industrial applications, out-door applications and so on, the need for more advanced polymeric materials and systems for coatings are important.

This field is vast and this presentation will highlight a few of the new polymeric systems being offered by the various manufacturers/suppliers in the particular field.

1. In keeping with industrial development activity, *The ERCA Group* offers advanced polymer solutions for a wide range of high-performance PU dispersions at low environmental impact. These products could be used for a sustainable production of low VOC synthetic materials for garments, shoes, bags, upholstery, sports goods and also suitable for coating of technical textiles.

   Applied in thin layers, they bring functionality to a material used in outdoor wear and also provide a please finish to denim, while protecting colour effects in thicker layers. PU coatings are the key for high quality synthetic materials that are used widely in footwear, garments, personal accessories, furniture upholstery and car upholstery. These ranges of PU systems can be used in different applications as direct coating, transfer coating (indirect), dipping and printing.

6.1.1.5 Other types of polymeric coated substrates

There are many types and qualities of polymeric coated fabrics and on other substrates. Once a coating system is set up, there are many other products that can be produced on the same equipment, although they may not come strictly under the category of artificial leather. Some coaters may also produce these types to boost their sales volumes as there is a big demand for these products or on the other hand, large volume coaters of standard artificial leather may not want to produce them due to their satisfactory profitable levels or lack of production time. For entrepreneurs with smaller coating systems, these will certainly produce valuable sales markets with high profits.

(1) Coated fabrics for safety wear

A huge market exists for this category. Some of the end applications are rain coats, capes, safety gloves, safety aprons, outdoor tents, firefighter's wear, shower curtains, tarpaulins, coated canvas and so on.

– **Weather products:** light backing fabrics like woven cottons coated on both sides with PVC and embossed with a smooth/plain roller. The finish can be either matt or glossy. For example, rain coats, capes and hats.

– **Safety aprons:** usually coated with PVC on both sides in various colours but generally in red or yellow colour, here, again a matt or glossy surface finish with a plain roller.

- **Outdoor tents:** light canvas coated with PVC with a water-proofing additive and a UV agent (optional). These coatings could be transparent also, if the surface colour/patterns are to be maintained.
- **Firefighter's wear:** these coatings have to be specially formulated with at least one highly effective heat-resistant system. Inside coating may differ and maybe designed for comfort. Usual colours would be yellow and with a smooth surface and should also be highly tear resistant. Naturally, the coatings must highly fireproof. A blend of polymers may be used to obtain the special properties required.
- **Shower curtains:** synthetic films in various thicknesses can be easily embossed with different patterns and coloured films can also be used for aesthetic values.
- **Tarpaulins:** coated with transparent or standard polymeric mixtures for heavy duty use. Coatings will probably contain UV and weather-proofing agents among others.

(2) Material for packaging

- At the end of each run of coatings, some polymeric mixtures of maybe different colours will remain as extra. Instead of considering them as waste and getting rid of them, they can be used to make some useful products. Newspaper producers and others who use reams of paper will always have 'ends' meaning leftovers from each run. These are normally rejected as waste and can be available free or at a very nominal cost. These coated lightly with the mixtures can ideally be used for packaging as wrapping paper, to make bags or other.

(3) Psychedelic coatings

- Polymeric mixtures of different colours can be used as combined coatings on substrates to produce ideal psychedelic patterns with a glossy finish for making handbags, fashion clothing, travel bags and so on.

(4) Coated fabrics with bio-fibre substrates

- The emergence of bio-fibre substrates like bamboo cloth, woven substrates with plant fibres coated with polymeric materials is a big boost for the small- to medium-sized handicrafts industries. Products like handbags, fashion bags, sun hats, floor mats outdoor applications are a lucrative business, especially for an entrepreneur and Asian countries are producing quality products with great export markets.

  Two interesting manufactures made in Sri Lanka are mentioned below:

  (a) *Pinatex* – an artificial leather made from a combination of natural and synthetic fibres. Commonly called pineapple leather, it is an imitation leather sold under the brand name 'Pinatex'. It is made from a blend of natural pineapple leaf fibres, thermoplastic polyester and coated with a petro-based polymer resin. Pinatex is available in many colours and finishes and used for the manufacture of shoes, bags, upholstery and other products.

(b) *Kantala* – a unique substrate made from a mix of natural fibres and bio-mass with polymeric coatings. Kantala's products are made with a distinctive blend of pineapple leaves, coconut shell and Hana (agave) rosette fibres. The name Kantala is derived from the Sanskrit name for the Hana plant. A small local company in Sri Lanka, inspired by a 300-year-old traditional weaving technology coupled with modern needs people's 'fashion' needs is making good quality handbags, marketing bags, mats, ornamental wall hangings, hats and other products with great export potential. Oeko-Tex 100 and ISO certified reactive dyes as well as natural dyes creates Kantala's characteristically unique products with vibrant colours and the variety of styles is wide enough to suit anyone's tastes.

(5) Viron – an alternative for footwear

Designed in Paris and made in Portugal, Viron is a plant-based material for footwear as an alternative for artificial leather. Their canvas uppers are made of 70% recycled cotton and 30% viscose.

(6) Adriano Di Marti's Desserto

According to information available (ref. nuvomagazine.com), the Adriano Di Marti Company has launched prototypes of gorgeous shoe and handbags under the brand name *Desserto*. These products are made from a hyper-sustainable leather alternative made from nopal (prickly pear), a cactus endemic to Mexico. It is especially eco-friendly because the cactus does not require much irrigation, relying mainly on rainwater and mineral-rich soil to grow organically and in great abundance.

Only mature leaves are used to create Desserto, which keeps the plant intact and ensures a new harvest in just 6–8 months. The resulting product is a soft, pliable and breathable fabric that is likely to expand beyond the fashion industry and into the realm of automobiles and home furnishings.

(7) Artificial leather with bamboo substrates

Bamboo fabrics like linen, bedsheets, towels and so have become very popular as they have special properties. The author suggests that foamed artificial leather made by an indirect/transfer coating process with suitable bamboo fabrics as the backing fabric laminated on to the foam layer instead of synthetic fabrics will have many end applications as an innovation. This will also contribute towards easing environmental concerns with less use of petro-based synthetic fabrics.

Generally, bamboo fabrics can be used for practically every application in which cotton is used. They are becoming very popular due to its notable beneficial attributes. For instance, bamboo fabric is highly breathable and also stretchier than cotton. Often these fabrics are thinner than their counterparts, while remaining similar or greater in tensile strength. Dyed fabrics in the colour of the polymeric coatings will enhance the aesthetic value of the final product.

(8) Polymeric coating on aramid-blended fabrics

Aramid fibres are a class of heat resistant and strong synthetic fibres. Their melting point is >500 C and the name is based on a blend of aromatic polyamides. They are used for speciality applications like aerospace, military, armour fabrics and ballistic composites and safety wear.

Manufacturers in China, India and Pakistan offer blended woven fabrics in a twill weave with blends of 93% meta-aramid, 5% para-aramid and 2% anti-static flame-retardant Nomex A.

6.2 Formulating with non-traditional fillers and stiffening/ softening agents

During the past few years, environmental concerns have been increasing most due to the use of petro-based products. This is very much so for the plastics industry, which have been dependent on refinement of crude oil for their basic monomers for the manufacture of synthetic polymers.

Additives serve important functions in a polymeric mixture and this is especially so in the manufacture of artificial leather. Constant research and development work is now coming up with alternatives form some components with bio-based ones. For detailed information, a reader may refer to the DG publication *Polymer Fillers and Stiffening Agents – Applications and Non-traditional Alternatives* (ISBN: 978-3-11-066989-3, 2020).

The following is based on making PVC artificial leather by a direct coating method. This manufacture will produce 'solid' PVC leather which will be made flexible as desired. Since the main objective is to show the use of non-traditional fillers and stiffening agents and the processing methods are vast, it is sufficient to present a basic processing formulation.

A general basic PVC mixture for coating will consist of the following:
- PVC polymer in powder form
- Filler polymer in powder form
- Plasticizers DOP, DIOP
- Filler 1 – egg shell powder
- Filler 2 – thickening agent (rice hulls/wheat hulls fine powder)
- Additives – lubricants, stabilizers
- Colourants – cadmium pigments
- Substrate – woven cotton fabric 54–108 in (137–275 cm)

The two polymers and the plasticizer will initially make a basic homogenous PVC paste. To make this mix suitable for coating, lubricants will help to create fluidity, while the stabilizers will be needed to prevent decomposition of the mixture. The pigments will give colour to the coatings and a desired colour may consist of two or three pigments mixed together. Since, the woven fabric is porous, the main PVC mix will have to be divided into two, with one (smaller batch) containing the thickening agent, which will prevent the PVC mix seeping through when the base coat is applied. The desired thickness of the artificial leather will be built with two or three coats with the larger standard batch.

Egg shells contain around 95% calcium carbonate and even a mix with standard calcium carbonate is acceptable. The rice hulls and wheat hulls contain around 20% and 8% silica, respectively and will function as an ideal barrier component. If these components are used in the form of ash (light grey), the silica content will be much higher. For example, rice hulls ash will contain around 70% silica. The actual amount of the thickening agent to be used on the base coat will depend on the thickness and weave structure of the woven cotton fabric used. Here, the services of an in-house laboratory will be very useful.

Basic formula (as an example)

Component	pbw	Base coat (kg)	Standard coat (kg)
PVC copolymer	100.00	50.00	90.00
Filler polymer	15.00	7.50	10.00
Plasticizer	50.00	25.00	70.00
Filler 1	15.00	7.50	12.00
Filler 2	0.10	0.05	–
Additives	3.00	1.50	2.40
Colourants	0.30	0.15	0.24
Blowing agent	–	–	–

6.2.1 Recommended processing method

A stock batch containing filler 1, colourants and a little of the plasticizer (wetting) is mixed thoroughly on a three-roll mill and made into a homogenous paste and kept aside. It is best if these stock batches are made in bulk from which to draw for each run. This means that a manufacturer will make batches of different colours for stock from which the required quantity can be drawn.

The base coat components without the thickening agent are put into a mixing vessel of a planetary mixer and mixed thoroughly to obtain a homogenous mix. The filler 2 (thickening agent) is now put into this mix and mixed again at slow speed at first and an increased speed to follow. A correct amount from the colour base stock is weighed and put into this mix. Here, ensure that the colour dispersion is satisfactory.

The standard coat can now be mixed with the colour stock being put in last and mixed thoroughly. In both mixing processes, it is likely that air bubbles will be generated due to the high shear forces being applied. A de-aeriation may have to be implemented. A simple direct coating process can take place by coating the cotton woven fabric first with the base coat and the then applying two or three coats of the standard mix to build up the final thickness desired. If necessary, make the middle coatings thickness slightly higher to compensate for height loss due to the pressure exerted by an embossing roller. If the embossing is done later when the coated fabric has cooled down, this extra coating height may not be necessary.

6.3 Graphene-induced coatings for automobile and industrial applications

Graphene has been around for some time but its actual potential due to its inherent super properties were discovered only recently. Two of the main beneficiaries have been the automobile industry and industrial applications as such. For example, because of its light weight, yet, properties of superior strength to metals, auto manufacturers are using graphene composites to make car bodies lighter. In industrial applications, the electronics and solar energy sectors are benefitting in a tremendous way.

6.3.1 Brief introduction to graphene

All of us are aware that common school-type pencils tips contain carbon in the form of graphite, the soft black 'writing-tip' and that the incredibly hard diamonds are basically carbon. Amazingly, both are allotropes of different forms of carbon. Another important fact is that both these radically different materials are made/formed of identical carbon atoms. Then, why is it that materials are different? The reason for this is that the atoms inside the two materials are arranged in different ways.

From the time graphene was discovered, it has become a very fascinating nanomaterial, thanks to its structure that imparts unique mechanical, electrical, thermal and very importantly very strong, yet very light properties. These factors have given scientists and researchers the impetus to look for different application sectors such as: catalysts and energy, Nano electronics, quantum physics and gives them the pathway to manufacturing of more advanced nanocomposites and biomaterials. Being a carbon source and having a well-ordered morphology, graphene is also being investigated for uses in another direction, That is in the fire retardancy of polymers, foams and textiles, often in combination with other flame retardant additives or nano-fillers.

However, in the last few years, scientists have found various other carbon allotropes with even more interesting properties. The key to these are formations of *crystal lattices*, a name for a solid internal crystalline structure. These lattices are made of lots of atoms arranged in a regular, endlessly repeating three-dimensional structure with invisible bonds between the atoms holding them together. Both diamonds and graphite have a three-dimensional structure, though completely different. In diamonds, the atoms are tightly bonded in three-dimensional tetrahedrons, whereas in graphite, atoms are tightly bonded in two-dimensional layers.

Graphene is a single layer of graphite. The remarkable thing is that its crystalline structure is *two-dimensional*. Just like in graphite, each layer of graphene is made up hexagonal 'rings' of carbon giving a honeycomb-like appearance. Just like in plastics resins, all of which are a little bit different and designed for varying applications, scientists are also finding different varieties of graphene. Some of the properties found to date are: almost completely transparent, extremely light, extremely strong and amazing conductors of electricity and heat.

An amazingly pure substance due to its simple orderly structure, graphene is based on tight regular atomic bonding. Since carbon is a non-metal, one might expect graphene to be one too but it behaves more like a metal and that has led scientists to classify it as a semi-metal or a semiconductor, meaning a metal between a conductor and insulator such as silicon or germanium.

Because of its unique properties and a major advantage of being super strong, especially for industrial applications, it is blended with materials like plastic resins, metals and others to make extra-strong composites. Take the example of applications for the automobile industry, the blending of graphene with polymeric coatings for better materials for auto interiors, energy-saving cars. Super light-weight yet, with very strong bodies, stronger composites for the building industry and so on – that is the kind of objectives one might expect for the future with this 'miracle' material.

6.3.2 Graphene-blended polymer coatings

PU is one of the main polymeric coatings used to make artificial leather. PU has important applications in coatings because of its outstanding properties such as high tensile strength, chemical and weathering resistance, good processability and also mechanical properties. However, since PU is an organic coating material, it is subject to deterioration due to degradation, mainly due to environmental factors like, UV light, water and oxygen. Any polymeric coating will tend to degrade when exposed to the influence of these factors.

UV light action will generally change the chemical structure of polymeric materials and this will affect both the physical properties such as loss of gloss, yellowing, blistering, cracking or other and mechanical properties such as loss of tensile strength, brittleness or other. The presence of water will promote the generation of hydroperoxides, which in turn will promote photodegradation of the PU.

According to researchers, in newly found technologies and coating systems, the addition of nanomaterials to polymeric coatings greatly improves the coatings because the nanomaterials really have high surface area to volume ratio, which gives exceptional properties to the new products.

6.3.3 Graphene-induced synthetic leather for improved weathering

This presentation discusses the addition of graphene nanoflakes to PU coatings in the production of artificial leather. Since weathering poses one of the biggest challenges for polymeric coated synthetic leather, researchers have found that the addition of graphene does in fact improve the resistance of the coatings, particularly, the surface coating. This factor is important as PU coated artificial leather is widely used in automobile, industrial, upholstery and other applications.

From end-use point of view, probably three of the most important ones are: automobile, aircraft and outdoor applications. With coatings of PUs, coaters can achieve an almost close feel and textures as compared to natural leather. PUs have important applications in coatings because of its outstanding properties such as high tensile strength, chemical and weathering resistance, good processability, mechanical properties and also high degree of flexibility. However, since PUs are organic coatings, they are subject to deterioration and degradation. Three of the main challenges for degradation are UV light, water and oxygen.

These critical factors cause degradation of polymeric materials when exposed to these environmental influences. Continued exposure to UV light will cause irradiation resulting in changes in both the chemical structure and physical properties. Coaters generally use a very thin protective transparent coating to protect the surface.

6.3.4 Graphene-induced synthetic leather to improve wear resistance

One of the challenges of using artificial leather is to improve the wear-resistance factor. Combined with weathering effects and constant usage, researchers have been carrying out many trials and tests to improve the wear resistance. With the advent and availability of graphene in recent years, some positive solutions have surfaced, one of which is as follows:

Chinese researchers have come up with recommendations which disclose a formula for producing synthetic leather from graphene-modified waterborne PUs. This is based on a coating process where a waterborne PU emulsion containing 5–15% of filler, 3–10% stearic acid and ammonium emulsion, 0.5–3.0% of a foaming agent, 0.001–1.0% graphene, 0.5–10% of water-based colour paste, 0.5–2.0% of ammonium hydroxide, 0.5–5.0 flatting agent and 1.0–3.0% of a thickener will make up the required coating mix.

The preparation of this mix will involve the mixing of the components, slow stirring at the start and then increased mixing speed to generate shear forces to obtain a good homogeneous mix. Since the colour is already in the form of a paste, it can be introduced directly into the main mix but the mixing must ensure thorough dispersion in the PU. The coating process will go through the normal coating and finishing procedures.

Bibliography

[1] Nurxat Nuraje, Shifath I. Khan, Heath Misak, Ramazan Asmatulu. article "The Addition of Graphene to Polymer Coatings for Improved Weathering." Open Access Article, August 2013
[2] Martin Melchiors, Michael Sonntag, Claus Kobusch, Eberhard Jurgens. "Recent Developments in aqueous two-component polyurethane (2K-PUR) coatings." www.elsevier.com-2000
[3] Fabric Funhouse-publication. "About Silicone Leather." https://fabricfunhouse.com/about-siliconeleather.
[4] Giulio Malucelli. Journal "The Role of Graphene in Flame Retardancy of Polymeric Materials: Recent Advances." Bentham Science, 2(1), 2018.
[5] Stephen Mraz. article "Mineral Fillers Improve Plastics." Machine Design, www.machinedesign.com/materials- Oct 2015.
[6] ERCA. publication "Textile Finishing and artificial leather." www.ercaps.com.
[7] Boris Hodakel. article "What is Bamboo Fabric: Properties, How Its Made and Where." https://sewport.com/fabrics-directory/bamboo-fabrics, Sept.2020.

Chapter 7
Machinery and equipment

7.1 Introduction

The manufacture of artificial leather entails sophisticated technology since its evolution from a simple polymeric coating on a fabric. At the beginning, polyvinyl chloride (PVC) was the most popular and used coating, and then the advent of polyurethanes and silicones improved the quality of artificial leather with very close resemblance to natural leather, while PVC remains the popular 'workhorse' among these materials even today.

Naturally, the use of these versatile polymeric coatings to take advantage of their full potentials required more sophisticated coating systems than the original simple direct coating machines. Due to intense research and development, a number of various coating systems are now available on the market, with *extrusion coating systems* from Davis Standard LLC probably being one of the latest techniques. Whichever coating system is used, the basic principles of coating remain the same, which are *direct* and *indirect* coatings.

Direct coated fabrics can produce both solid- and foam-coated fabrics, while indirect coatings will generally produce highly flexible foam-coated artificial leather, some very close to natural leather. In this process, since embossed release paper is used, the number of times this paper can be used depends on the proper 'handling' of it in the coating process, to a certain degree on the set-up, accuracy of the forward and reverse movements of the paper, and to a certain degree on the alignment and accuracy of the coating line to prevent tearing, surface scorching due to excess heat and other things. Since these papers are expensive, this could be an important aspect of costs.

Another area of importance on a coating line is the heating ovens, which can be one/two for direct coatings and several for indirect coatings. The number and lengths of ovens required for a particular coating system generally depend on several factors such as the type of polymeric mixes, thickness of coatings, the speeds required, evaporating rates and solid to blow ratios for foams. A coater will, in all probability, use temperature gradients in the ovens of his coating system and this means very effective and efficient temperature controls for all ovens,

There are many manufacturers and suppliers of machinery and equipment, and this chapter will deal with selected coating systems suppliers to demonstrate the versatility of the range of machinery available for coaters. Some may opt for custom-made systems, especially when specialized materials have to be produced.

Coating heads play an important role also, to ensure a smooth and even coating thickness, and these are enhanced by automatic pre-settings in modern systems. There are possible fire hazards as the coated material goes through the curing ovens

https://doi.org/10.1515/9783110716542-007

and at the embossing station, where an independent vertically moveable heating arrangement will be used to heat the coated surface sufficiently to enable a good emboss. In the case of indirect coating, the emboss patterns are obtained by the embossed release paper. Whatever coating systems are used, the exhaust systems must be very effective.

The basics of a coating system will consist of the following:

7.1.1 Direct coating

- Fabric supply station
- Coating head
- Heating oven
- Exhaust system
- Embossing system
- Take-up station

7.1.2 Indirect or transfer coating

- Fabric supply station
- Release paper supply station
- Coating head
- Laminating station
- Heating oven no. 1
- Heating oven no. 2
- Exhaust system on ovens
- Embossing station
- Coated fabric take-up station
- Release paper take-up

Artificial leather is manufactured by several coatings on a base fabric or the lamination of a knitted fabric on foam and a coater, depending on the size of the production operation will decide whether he/she will invest in standard machinery where the coatings are done on a forward/reverse basis. Some may even do the embossing and finishing operations on separate units when the coated materials have cooled down. This is an advantage when foamed materials are produced, as it will prevent crushing the hot foam when pressure is applied by the embossing/printing rollers.

For large volume producers, one-pass coating systems are available but the initial investment will be heavy. To operate these systems a few experienced operators will be needed. Some of the coating systems are:

- Doctor-knife coaters
- Gravure
- Reverse roll
- Slot die
- Rod coater
- Smooth roll
- Air-knife
- Curtain coater
- Dip and squeeze
- Knife over roll
- Hot melt

From these different types of coaters, a manufacturer can select the best system for his/her operation. In addition to this, other machinery like triple-roll mills and mixers are needed. It is customary when purchasing these machineries for the suppliers to install and train operating personnel and assist them with the initial trials.

7.2 Coating lines

The following demonstrates two types of coating lines:

7.2.1 Indirect coating line

This coating line is offered by Crown Machinery Company Ltd. in Taiwan and has the following features:
 Key features:
- Tailored design equipment width and speed according to customer needs
- Multi-layer coating and special laminating capability allowing instant peeling process
- Special knife-blade design delivering accurate coating thickness
- Accurate transcribing of the release paper patterns
- Advanced dryer design providing high drying efficiency and uniform drying
- High efficiency, energy conservation and superior safety
- Excellent tension control
- Non-stop semi-automatic splicing mechanism allowing long-term continuous production runs
- Robust structure and body design ensuring long usable life.

Figure 7.1 shows a photo of the artificial coating line:

Figure 7.1: Photo courtesy of Crown Machinery Company Ltd. Taiwan.

7.2.2 Uniroll SP-Rollmac SPA Italy

It offers a coating line with traditional coating head which is available in two models. TF model is for simple reliable thickness coating and TM is for thickness and 'on air' coatings. These coating lines have these special features:

Uniroll SP is a sturdy machine to guarantee very thin coating thicknesses. It features a two-position blade holder for a quick change by manual rotation between the two different knives. The blade holder is mounted on two side turrets which can be shifted on their horizontal axis in order to coat by knife over roller or 'on-air'. The shifting is manual by hand wheels or, as an option, motorized.

The adjustment of the working thickness (clearance) is motorized and independent for the two sides with quote visualization for digital display located on the control panel. The machine includes an electronic panel with PLC mounted on the machine to manage the various functions.

7.3 Polymeric mixtures for coatings

One of the most important aspects of manufacturing artificial leather by coating polymeric mixtures on textiles is achieving a final very homogenous mixture to ensure smooth coating layers, uniform colour distribution and even coating thickness across the full width of the substrate being coated. Challenges like *lumps*, *poor colour distribution*, *air bubbles*, *viscosity variations*, *uneven foam cell structures* and *foreign matter particles* will hinder achieving good and smooth coatings.

There is no doubt that the accuracy and smooth functioning of a coating line plays a big part in the success of manufacturing artificial leather of good quality. However, the preparation of high-quality polymeric coating mixtures must take precedence. Therefore, a coater must select the best mixing, blending and dispensing machines to suit a particular operation. Generally, the polymeric mixtures prepared for coatings will probably be similar for both direct and indirect coating operations, with the difference probably being the use of foaming agents.

Since formulations will contain some components like plasticizers, solvents, fillers, blowing agents, thickening agents, cell openers, pigments and other additives, the process of mixing cannot be done in one operation where all components are put into a large vessel and mixed together. For this purpose, there are various types of mixers which can mix the components as 'groups' and then brought together in one large vessel for the final mixing operation.

The following are some typical mixing and blending machines that can be used:
- Triple-roll mills
- Planetary mixers
- Ribbon blenders
- Ball mills
- Vacuum blenders
- High-shear mixers
- Two-roll mills

Due to constant development in the engineering sector, these machines undergo periodic improvements, and coaters have the privilege of selecting these new machines on the basis of investment *vs* benefits for upgrading their manufacturing processes.

Of the above, the two main mixers needed are the triple-roll mills and the planetary mixers. The types and number of mixers will depend on the producers of polymeric mixtures or a coater on a production floor. For example, take the case of manufacturing PVC-coated artificial leather using a basic direct coating method.

Polymeric mixture formula – example
Copolymer
Plasticizer
Filler
Pigment
Stabilizer
Lubricant
Thickening agent
Additives
Blowing agent (optional)

Using this formula, a manufacturer of PVC artificial leather would mix the filler, pigment and a little plasticizer on a triple-roll mill and make a paste. This could be a batch sufficient for a single coating run or could be larger batches for stock to be used later. The copolymer and the plasticizer will be put into a mixing vessel of a planetary mixer, and when a fairly homogenous mix is achieved, the stabilizer, lubricant and additives can be added and mixed thoroughly. Then the correct amount of the filler/pigment paste can be added and mixed again. From this, the mixtures for the base coat (with thickening agent) and skin/top coat (without filler) can be mixed as separate batches. If a final transparent protective surface coat is desired, this can be done without filler, pigment and thickening agent.

7.3.1 Triple-roll mills

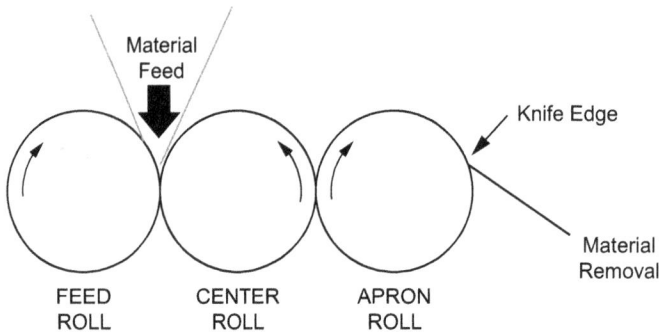

Figure 7.2: Triple-roll mill – modified by the author.

A triple-roll mill or a three-roll mill, as shown in Figure 7.2, is a mixing machine that uses shear forces created by three horizontally positioned rolls, rotating in opposite directions at different speeds relative to each other in order to mix, refine and dispense a homogenous mass of any viscous materials fed into it. These machines may vary from a laboratory table-top mill to small-size, small-volume mills to large-volume industrial types.

A triple-roll mill consists of three adjacent rollers called the *feed roll*, *centre roll* and *apron roll* which rotates at progressively higher speeds. Material usually in the form of a paste is fed between the feed roll and the centre roll. Due to the narrowing of the gaps (space) between the rolls, most of the mix or paste fed in will initially remain in the feed area. The part of the feed material that makes it through the first in-running nip will experience very high shear force due to the different rotating speeds of the two rollers. Upon exiting, the material that remains on the centre roll moves through the second nip between the centre roll and apron roll. The nip-gap between the centre roll and the apron roll, being the smallest, will create higher shear forces which will produce a very refined homogeneous material A knife blade then 'scrapes' this material off the apron roll and down the apron roll to be collected into a vessel. Depending on the degree of refinement needed, this milling operation can be repeated several times to maximize dispersion.

The gaps between the rolls can be mechanically or hydraulically adjusted and maintained. Typically at the beginning, the gap aperture is greater than the particle sizes of the material being fed in, and in some operations the gaps are progressively decreased to achieve the level of dispersion desired. To counter heat built up due to shear forces being applied and also friction, the rollers are cooled by water internally.

A typical example in practice during an artificial leather manufacturing process is the mixing of a filler/fillers with pigments to obtain a homogeneous mass, eliminating globules due to unmixed pigments, coarse filler particles and ensuring a smooth and even distribution of colour. This is a better and more effective way of even colour distribution in a final polymeric mixture used for coatings. The other alternative is to add the pigments/dyes directly into the polymeric components being mixed, for example, on a planetary mixer.

Small bench type triple-roll mills are used for laboratory work, while larger models will be used for small batch productions and pilot project works. Main advantages of this process are that it allows high-viscosity pastes to be milled, and the high surface contact with the cooled rollers allows the temperature to remain very low, despite the dispersion work being done at high speeds. However, one disadvantage may be the partial loss of solvents, if used, instead of a plasticizer or other things.

7.3.2 Planetary mixers

In the coatings industry, in general, the focus has been changing from conventional low-solid, solvent-based formulations to waterborne systems, high-solid coatings and other things in keeping with low volatile organic compound technologies. This important environmentally responsible shift coincides with the exciting new developments of more advanced and versatile materials in response to industrial and customer needs.

Whatever new materials are used in the coating industry, especially in the preparation of polymeric coatings, the most important factor remains in the blending and proper mixing to achieve high-quality mixtures. The degree of mixing and dispersion applied will invariably affect its colour, gloss, stability, flexibility, adhesion properties, curing rates, weatherability, and other performance characteristics required by the end applications. From a business angle, it makes sense to carry out periodic assessments of this sector of operations and even well-established mixing processes can benefit from new, more advanced systems. This is especially true for large-volume manufacturers of artificial leather.

A basic planetary mixer would consist of a sturdy stand, two removable blades for mixing and a large stainless steel bowl. Some models may have the mixing head and blades on a fixed base with the mixing bowl being able to be lowered down, or a removable mixing bowl secured at the base of the mixer and the mixing head and blades being able to be tilted up or raised. Planetary mixers will have different mixing speeds and will generally work at high speeds. Newer models are called double planetary mixers with more efficient mixing processes. There are varieties of new models from different manufacturers.

7.3.3 High-speed dispersers

The high-speed disperser, previously called a high-speed dissolver, is a standard workhorse in the coating industry. The coating industry means a very large spectrum of activity and coatings for artificial leather is one of them. These machines are economical and are a relatively simple piece of machinery; its prime purpose is to incorporate powders into liquids and break down loose agglomerates to produce an acceptable level of dispersion prior to milling. Running at high speeds, the open disc blade on a shaft creates vigorous turbulent flow within a low-viscosity batch. It will also generate a characteristic vortex into which dry ingredients can be added for quick mixing. The disperser blade may be located on- or off-centre, depending on the depth of the vortex. A centre blade will create a deeper vortex, while a blade at off-centre position will create a smaller vortex and reduce air entrapment. In coatings for solid polymeric mixtures, the removal of air is an advantage, while for foam coatings, the build-up of air bubbles will be an advantage.

7.4 Equipment

The manufacture of artificial leather is quite exciting with good production challenges. To ensure a smooth operation and good production flow, sound workable equipment and floor designs are also of primary importance. A manufacturing plant may consist of several coating lines and the number of mixing, blending and other auxiliary equipment will depend on the volume of production. The following are some of the basic equipment needed:

7.4.1 Exhaust/ventilation systems

Since these manufactures will generate toxic gases from PVC and polyurethanes, it is essential that efficient exhaust and ventilation systems are in place. Generally, each heating/curing ovens will have exhausts on top of each unit, but it is necessary to have an overall exhaust system possibly to cover at least some of the areas. Ventilation is also important and the overall efficiency can be verified when an independent organization is called in to test the air quality against standards required. Naturally, a manufacturer will test it for himself/herself first before a professional opinion is sought with a certification, if needed, according to industrial standards.

7.4.2 Overhead mobile lifting system

Since the production floor will be dealing with materials like heavy release paper rolls, large rolls of fabrics, large rolls of coated fabrics and embossing rollers, an overhead mobile lifting system covering all coating lines and others, wherever necessary, will be a big advantage. In certain coating lines, where the embossing is done in one operation, the embossing rollers may be mounted on a frame above the slowly moving coated cloth and a selected roller can be lowered or lifted up electronically. This could apply to embossing machines where the embossing is done at a later stage.

7.4.3 Fork lifts

A couple of fork lifts will be needed to transport materials from storage to their respective operational machinery. If an overhead lifting system is not available, fork lifts could also perform the above tasks. Platform trolleys and other in-house transporting devices would also be useful.

7.4.4 Weighing machines

Electronic ones would be the best. For an in-house laboratory, weighing machines from 10g to at least 1 kg. For the production floor, industrial types include a few platform weighing machines also.

7.4.5 Air compressors

It would be best to house them in a small room, inter-connected and drawn from an accumulator tank for any pneumatic or other operations. Compressed air supply will have a drop in volume over distance and should have a compensating factor when calculating required volumes at final points of draw.

7.4.6 Measuring tables or systems

The coated fabrics will be in large rolls. All coatings will have an 'edge' which has to be trimmed. Some production lines will have an automatic trimming system off the coating lines before being rolled up on a horizontal axle. An alternative is to trim the edges on a post-coating basis by another machine before being inspected and packed into standard lengths. An entrepreneur, instead of investing in an electronic trimming machine, would in all probability use a long horizontal measuring table with a simple trimming device on both sides.

7.4.7 Other equipment

Packed and QC-approved coated rolls have to be moved to the finished goods/shipping area. Here, there should be a well-organized racking system to stack the finished goods efficiently and for easy identification. Special care should be taken with foamed fabrics so as to not crush the coated foam.

7.4.8 Tools

Standard range of tools are controlled either from an in-house workshop or a tool room. Special tools related to the coating lines, mixing machines and maintenance work may also be required.

7.4.9 Collection bins

There will be generation of wastes from the mixing, coating and coated fabrics and QC areas, and suitable large bins or containers will be needed, preferably on wheels for easy mobility. Since different types of wastes are involved, these collection vessels must be chosen carefully.

    The above recommendations are given on the basis of a standard artificial manufacturing operation but the needs for a fully automatic manufacturing operation may be different in some aspects.

Bibliography

[1] Banaszek, Christine. article on "High Speed Mixers for Coatings and Inks". February 1, 2015, PCI Paint & Coatings Industry.

[2] Engineering 360. Planetary Mixers Selection Guide. www.globalspec.com/learnmore.

[3] ROSS. Industrial Blender and Mixing articles. www.mixers.com/resources/articles.

[4] Harnby, N, Edwards, M.F., Nienow, A. W. (1997). "Mixing in the Process Industries", Butterworth, Heinemann, 128–130

[5] Davis Standard. Liquid Coating & Laminating Tips. https://davis-standard.com/custom_blog.

.

Chapter 8
Coating systems

8.1 Introduction

There are many different coating systems but basically is made up of a coating head/material feed unit, laminating unit, release paper feed unit, heating ovens, take-up units and embossing/finishing stations, which can be separate from the main coating system. The dimensions of a system like width, length and type will depend mostly on the width of the fabrics and type of polymeric being used. The heating/curing ovens play a major role on a production run with basic functions of producing a uniform foam layer and a total cure of the whole coated material. Multiple ovens will have gradually increasing temperature gradients. Since fire hazards are always a possibility, the oven temperatures must be well controlled at all times, especially the extra heating system to be lowered for surface heating before embossing. If too close to the surface, it will catch fire.

The manufacture of artificial leather requires a good knowledge of technology of polymers, substrates to be used and methods of application. There are a number of polymeric materials that can be coated on various fabrics, substrates and products. Polymeric coatings cover a vast area of activity in industry and manufacture of artificial leather can be included as a major activity under its umbrella. Artificial leather, originally used for upholstery, has developed very fast over the years and now covers applications in major sectors in consumer, industrial, automotive, building industry, transport, travel, space travel, fashions and so on.

Coating systems technologies have also evolved rapidly to meet the needs of the coatings for artificial leather beginning with polyvinyl chloride (PVC) to polyurethanes (PUR) to one of the latest silicones. Although there are other types, these three plays a major role and meets the needs of 90% of the artificial industry. Artificial leather qualities and textures have improved tremendously with very close resemblance to natural leather so much so that sometimes even experts may find it difficult to tell the difference.

Common substrates used for coatings are cotton, synthetic materials and bio-based ones made from plant fibre. Fabrics from mixed fibres can also be used depending on the end applications. These can be either woven, non-woven or knitted fabrics, with the knitted ones being used for foamed coatings using indirect/transfer coating systems. In direct coating systems, a fabric is 'directly' coated onto the surface with a base coat, built up with one or two solid or foam coats, followed by a final (skin) coat which can be embossed and finished with printing where desired.

Although there are many types of polymeric coatings that are possible with a formulation based on a single or a combination of different polymers, this presentation is based on PVC, PURs and silicones as these materials will meet almost any

https://doi.org/10.1515/9783110716542-008

application required. PUR, and especially silicones, with emerging technologies are further providing exciting possibilities for the artificial industry. The recent discovery of graphene and its possibilities in combination with polymers for the coating sector will no doubt enhance it tremendously.

8.1.1 Common polymers for polymeric coatings

Polymeric coatings on fabrics, especially in artificial leather, are used to protect the substrate, and with different finishes enhance its surface aesthetics. They can be used as a single polymer or in combination for preparing polymeric mixtures as coatings for artificial leather. There are many polymers that can be used but for artificial leather the most common ones in industry are as follows:
– PVC
– PUR
– Silicones (Si)

Each of the above will provide excellent performance-driven, protective characteristics depending on the desired end application. Of these, the most common ones used for making synthetic artificial leather are PVC, PUR and silicone.

8.1.1.1 Rheology of polymer coatings
This is an important aspect for polymeric coating processes. Polymeric coatings generally will be in the form of pastes, liquids or 'soft solids', and in coatings the flow properties are of primary importance. Rheology is the study and evaluation of changes that will take place when a force or forces are applied to matter and in this case polymers. Rheological measurements, therefore, when analysed will show how each polymer or a combination of polymers will behave during processing.

This information will greatly help in understanding the behaviour of coated products under different conditions of a coating process enabling good control of coating parameters to achieve quality products. For example, a coating material must be stable during storage, easy to apply whatever coating method is used and ideally an excellent final surface finish. As a result, rheological evaluation and knowledge of polymeric behaviour will determine to a great extent coating parameters and will also become a key factor in monitoring and development of coatings.

Some of the terms used in rheological analysis are as follows:

Shear stress/strain: while shear stress is defined as a force being applied to a material, the resulting deformation and the extent of this is defined by shear strain.

Shear rate: shear rate is dependent on time and the shear rate that materials are subjected to can change significantly during a coating process.

Viscosity: in general terms, viscosity is accepted as 'thick' or 'thin' and can be defined as the shear stress divided by the shear rate. Fluids like water will have low viscosity as there will be hardly any resistance to flow when a force is applied. On the other hand, a PVC paste could have a thicker viscosity which could be lowered by an additive like a plasticizer. In practice, it has been found that PUR in storage, for example, in drums will increase in viscosity and has to be stirred well before use. Viscosity is a function of four parameters: shear rate, pressure, time and temperature.

Thixotrophy: this is a term which describes a property in a material that changes its viscosity with time. Shearing or shear forces will decrease viscosity but when removed will slowly tend to regain its original viscosity.

Viscoelasticity: materials have viscoelastic properties when they exhibit both viscous and elastic properties. Unlike standard foams which are two dimensional, viscoelastic materials are based on four-dimensional parameters such as density, time, hardness and pressure.

Normal force: during a coating process, high forces due to stress at the coating knife will tend to increase resistance and push the coating blade away and variation in normal forces will result in uneven coating thicknesses across the substrate or backing fabric.

8.1.1.2 Application of polymer rheology

Applied forces during a coating process will make polymer fluids to behave in a viscous or elastic manner. In order to ensure an even coating thickness on the surface of the backing material, it is important to know the rheological aspects of a polymeric coating material, whether it is of a single polymer or of a combination of polymers. The rheological structure of the coating mass is the key to successful processing. This means the processing equipment used must be able to handle the parameters being set for a coating process or the coating material must be modified to suit the machinery limitations.

The aim when formulating a coating material is to achieve its final end application needs. If modification of a coating material is required for enhancing a coating process, it has to be done in such a way so as not to hinder the performance of the final product. Due to the sensitivity of behavioural changes of coating materials, rheology may be considered as the best technique to control and improve a process.

It is to be accepted that material batches from same grades may behave differently during a coating process. While this may not be very significant in thin coatings, in thicker coatings these variations may pose problems. The properties of a polymer melt are sensitive to changes in the polymer structure. For example, a small amount of a high-molecular-weight (MW) polymer can change the processing behaviour dramatically and so does the melt rheology.

Three of the basic structure parameters defining the rheology of polymer melts are MW, MW distribution (MWD) and branching. Whereas an increase in MW will cause an increase in viscosity, changing the MWD mainly affects the elasticity of a melt.

8.1.1.3 Rheology of PVC plastisols for coatings

A plastisol can be defined as a suspension of PVC particles and additives in a liquid phase, composed of a plasticizer or a combination of plasticizers. PVC plastisols can be formulated for use in many different processes to produce various end products. One of biggest applications is for production of artificial leather using both the direct and indirect coating systems. In many PVC-based formulations for coatings, they may contain more than one PVC polymer. The PVC mass generally could start as a paste with the addition of fillers which will increase its viscosity, and then the addition of a plasticizer will reduce the viscosity and the shear action of mixing will produce a homogenous mass. For additional fluidity and ease of flow, a lubricant can be added. In a coating process, a certain amount of mass viscosity is needed and formulated accordingly. The mixing process starting at slow speed and then increasing gradually will create air bubbles due to the high shear action. For direct solid-coated leather, these air bubbles have to be removed before coating, whereas for foamed PVC this would be a plus.

8.1.1.4 Polymer rheology-extrusion coating

Although PVC, PUR and silicones are at the forefront of the artificial leather industry, polyethylene-coated fabrics also forms a large area of demand for various end applications. With extrusion coating processes becoming popular, it is best to have even a basic understanding of the rheology as related to extrusion coating machinery.

Other polymers, although not strictly used in making artificial leather, are used for protective coatings, for example, canvas and outdoor materials. In extrusion coating technology, T-shaped wide flat dies technology has emerged as the premium technology to handle monolayer and multilayer coatings.

Here, the author recommends the addition of 10–15% rice hulls ash (around 70% silica) to the polymer melt which will give greater protection when exposed to the elements, with the silica acting as a strengthening and effective moisture barrier.

Machinery manufacturers for these processes have to find innovative solutions to fine balance the constant trending of artificial leather products which are improved quality, specialized needs, especially for the automobile and transport requirements, faster production volumes and to be cost-effective. If the process equipment and production results are satisfactory, then the initial investment of a complete extrusion coating line may not be an issue.

This means that machinery manufacturers when designing and effecting innovations to meet the new challenges should be aware of the range of polymers to be processed. Higher output rates means higher volume flow of a melt exiting from a die, which will affect melt properties. Basic melt measurements do not quantify effects on polymer flow and behaviour during a process. Therefore, additional data are necessary for improving an existing process line or for designing a completely new one.

In an extrusion coating process, the polymer mass is put into a hopper of an extruder having a die with a typical adjustable die opening, which should be able to accommodate the width of the substrate being used, depending on the capacity of the extruder to give sufficient volume/flow.

8.1.2 Advanced coating polymers

Current sophisticated markets for artificial leather with ever-increasing end applications have to meet more complex properties like antimicrobial, fire-retardant, metallic, electrically conductive and other attributes, depending on the individual end application. Today's advanced coating techniques also include phase changing coatings and the use of increasingly advanced fabrics. In an increasing demanding market for technical conducive fabrics, these fabrics will soon become an essential facet of fabric coatings.

8.1.3 Some applications for coated fabrics

Coated fabrics have a very wide range of applications. It is too numerous to identify each product and some of the main categories are as follows:
- Upholstery
- Footwear
- Sports goods
- Fashion wear/goods
- Protective wear
- Tarpaulins
- Material for outdoor applications
- Medical applications
- Packaging
- Construction
- Geotextiles
- Aviation
- Agriculture
- Other

Here, it should be mentioned that there are other polymeric coated items made with paper, synthetic films, materials made with biomass fibres and others as substrates that have a great demand in the market, although they cannot be categorized under 'artificial leather' or under coated fabrics. These can be easily produced on the same coating systems used for making artificial leather, especially, basic systems used for direct coating. For example, a coater can produce coated paper with left-over polymeric mixes for packaging, decorative packs or other, embossed synthetic materials for shower curtains and so on. Since the probable left-over mixes will be of different colours, these can be used to produce attractive psychedelic finish.

8.2 Coating technology in brief

Manufacturing artificial leather is a process in which a coating material is applied directly or indirectly to one or both surfaces of a chosen fabric. In the case of indirect or transfer coating, the first layer is applied onto an embossed release paper to form a 'skin', followed by other layers to obtain the desired thickness and then the fabric is laminated onto the last layer. Generally, the polymeric mixtures are viscous and a 'doctor knife' or an 'air knife' across the full width of the slowly moving fabric will control the desired coating thickness of each coating layer by a pre-set aperture at the coating head/unit.

The final thickness desired is built up by two or more coatings with each layer going through a curing/heating oven and semi-cured to provide good adhesion for the following layers. The final layer can be fully cured and as it emerges out of the final oven, embossed, printed or other as desired. In both direct and indirect coating processes, the skin layer is a specially formulated one applied for the enhancement of the aesthetic or the technical aspects of the product. Some coaters will apply a very thin transparent coat as a protective one. This coat must be highly transparent and flexible so as not to hinder any of the qualities of the surface finishes.

In processes where woven fabrics are used, for example, in direct coating, the first layer (base layer) will contain an additive to increase viscosity to prevent the polymeric mix seeping through the open weave structure of the fabric. In the case of indirect or transfer coating, this problem will not arise, although the knitted fabrics will probably have a more open structure, since they will be laminated onto the last coating. Here, a coater has to be careful to achieve only sufficient lamination pressure to obtain good adhesion to the foam coat. Too much lamination pressure will make the knitted fabric 'sink' into the wet foam layer, thus, affecting flexibility of the artificial leather. Another point of possible importance is to ensure absolute horizontal lamination across the full width of the foam layer and not at a 'tangential angle'. Too much lamination pressure will also result in 'crushing' the foam, destroying cell structures and, thus, softness of the final product.

In direct coating processes where the embossing is done online, final coated thickness reduction is possible due to excess embossing pressure on the hot/warm polymeric coating. Two possible solutions for this are to increase the overall coating thickness slightly to compensate for 'embossing loss' or differ the embossing/finishing process for later when the coated material has cooled down. Here, a slight warming of the top surface will be needed in any case to soften the surface in order to obtain a good emboss.

When using woven fabrics for direct coating, there will be reduction in the final width (weft) of the coated material. Woven fabrics will generally come in large-length rolls and preferably without any joints which will create rejects. General practice in preparation of the woven material prior to coating is to have a 'lead' cloth or other attached to one end of the roll which will be mounted on the take-up station and treaded through the entire coating line length and secured at the feed end. An operator will then wind the material at slow speed and under tension to spread the material weft-wise. During this operation, the material under tension will undergo reduction in width, for example, a 108 cm width will become 106 cm which will be the surface width available for coating. During this winding operation, the operator will look for surface flaws like knots, fluff, any threads sticking out and so on and remove them. Some coaters may carry out this operation on a simple separate machine which will save some time on the coating line.

In fabrics used for both processes – direct and indirect – there will be 'edges' across the wefts, generally between 0.25 and 0.5 in. The basic purposes of these are to 'hold' the weaves/knits in place and to prevent unravelling. The coatings will be within these borders and these edges have to be trimmed off either online or later during packing.

8.3 Coated textile materials

Coating on textiles can be both exciting and challenging. While polymeric solid coatings may be considered as quite straightforward manufactures, production of foamed coatings will need a good knowledge of technology, coating and finishing techniques. The production of surface finishes like matt, gloss, high gloss, metallic, silver and gold will also need special attention. The application of final top surface protective coatings will need formulations to meet high transparency, flexibility and at least durability.

The selection of processing equipment and machinery will also play an important part in the manufacture of artificial leather and how well they perform. Here, the two processing areas of primary importance are the coating head unit and the heating/curing systems. While the coated material goes through the curing ovens, they must ensure even heat application on the total surface of the coating which will be in slow motion. On a two- or three-oven system, the temperatures will be set on a gradient starting at lower heat, increasing gradually. For solid coatings reversing of the base coat and winding for a second or third coat will not pose any problems. However, in foamed coatings, during rewinding for additional coats, problems may arise with 'crushed' foams if the pressure is too much.

When first oil cloth was made with a single coating of linseed oil and subsequently the number of coats was increased to improve the product, this procedure may be the predecessor of today's multi-layer coated fabrics. Multi-layered coated fabrics, coated on one side or both sides, will definitely have more special properties than a single type of polymer-coated material but these are for specialized applications like safety wear and space travel. For artificial leather in general, for example, for upholstery coatings will be on one side of a chosen fabric.

Some types of polymeric layers can be, for example, PVC, PUR and silicone, which are the popular ones. Coatings can be applied to textiles either directly or indirectly using release paper. The constant development of coating techniques has resulted in new achievements like the application of nanoporous polymers to a textile substrate. One of the uses of these products is increasing its importance especially in the clothing industry. Their importance is highlighted with materials for the protective wear sector, where these materials meet all requirements of technical standards.

Straightforward coated materials would hardly meet the exacting properties needed for high-profile materials for applications like space travel, fire-fighters, outdoor products and many other things. Therefore, the birth of multi-layered polymeric-coated textiles may be classified as composite materials. These could be one side or coatings on both sides and could be layered with different polymers to reach the final objective of desired properties. Here, three main areas are to be considered: environmental conditions, heat and cold with flexibility and wear comfort. To enhance coatings, backing materials or textiles are treated before coatings are applied.

Multi-coated fabrics can be produced as follows:
- By laminating polymer layers to a textile surface on one side or both sides
- By directly applying several polymeric coatings to a textile material on one side
- By indirectly or transfer coating a polymeric coating to a textile material, where the fabric is laminated onto the last coat

8.3.1 Laminated fabrics with polyurethane coatings

Laminated fabrics with PUR are multi-layered composite materials. Fabrics used may comprise two or more textiles, and PUR formulations used may differ for each layer. All components comprising the final product are carefully designed to meet the required standards of its end application. Multi-layered composite fabrics coated with specially formulated polymerics will have many advantages over standard coated textiles, since they are more durable, stronger, have high protection against metrological effects, have good sweat permeability, more resistant to abrasion and load and will have less anisotropic properties.

When woven fabrics are used as the base textile, the quality of the final product will be influenced to a great extent on the type of weave, warp and weft density, yarn count and the angle of the straight line under which a load acts in relation to the warp and weft direction. A higher breaking strength is expected in the warp direction (length) than in the weft direction (width). Lamination can be between fabrics or a fabric and other materials.

8.3.2 Polymeric coatings on textile materials

Polymer coatings on fabrics provide new properties converting them into better products, one of which is artificial leather. With constant research and development taking place on both natural and synthetic fibres and advances in newer polymers or their combinations, the qualities of artificial leather being produced are advancing at a rapid pace. As the demand for artificial leather as a substitute for natural leather increases, manufacturers have to also meet newer challenges, especially in the automobile sector as well as in other sectors where exact properties are being sought for different applications with aesthetic values also being an important factor.

Since artificial leather or synthetic leather is a combination of polymeric coatings on a substrate, both have to contribute their best possible properties to produce a suitable leather to meet the requirements of specific applications. The fast developing bio-fibre textiles as substrates are creating exciting challenges to coaters.

In order to achieve coated materials suitable for its intended end applications, it is important that the properties of the substrates used are able to successfully combine with the coating/coatings to achieve desired qualities. The following properties of substrates used are of special importance:
- Good mechanical properties such as elasticity and tensile strength
- Type of yarn used: if the yarn used has short hairs sticking out, it will promote good adhesion for thicker coatings, while these will stick out in thinner coatings and be a problem
- Dimensional stability: fabrics undergo tension on a coating machine and must, therefore, be dimensionally stable to overcome these forces
- Adhesion, absorption: in indirect coatings, this is not a problem as the backing fabric is 'laid down' or laminated onto the last coating. In direct coatings, where mostly woven fabrics are used, the inherent open weave will tend to allow the coating to seep through.
- Pre-treatment: additives such as softeners and dyes can negatively affect the subsequent strength of a substrate, whereas other additives such as thermal stabilizers and antibacterial agents can improve the quality of the final product.

– Thermal stability: some polymeric coatings will need high temperatures to cure them and, thus, substrates used for these coatings will have to withstand these high temperatures to negate the possibility of poor tear strength.
– Uniformity of substrates: fabrics/textiles selected for coatings must have uniform thickness and weight across the weft and along the warp. Dimensional stability plays a key role, although a certain percentage of width loss is to be expected due to tensioning on the coating machine.

To make artificial leather, generally the three basic polymers used are PVC, PUR and silicones. There are other things that can be used as mentioned earlier in the presentation. For products that need to meet high technical specifications, some of these polymers can be combined or used as different layers. If in combination, compatibility is important, whereas used as different layers on the same base fabric, adhesion to one another becomes important. To promote/improve these characteristics, there are several effective additives that can be used.

Although there are different methods of applying polymeric materials to fabrics, the basic principles of applications are based on either direct or indirect coating (transfer coating). In direct coating, the polymer is directly applied on the fabric and several coatings may be done to achieve the final thickness required. In indirect coating, the polymer is first coated onto an embossed or plain release paper coated with silicone for easy release. These release papers come in very large rolls and are expensive. Once they are mounted on a coating machine, they will be used for coatings and will go from the feed-end through the ovens and be wound up and come back again to the feed-end for subsequent coatings. The number of times these papers can be used will depend on how well it is used in the production process. Some hazards are scorching, tearing, non-alignment or other. They can be purchased based on their weight, for example, grams per square metre and thicker ones will last longer. Generally, it is accepted that they can be used 8 to 10 times.

8.3.3 Foamed coatings for artificial leather

Both PVC and PUR can be used for foamed coatings, using either a direct coating or indirect (transfer coating) system. The coatings will be formulated on the base polymer and additives in a consistency suitable for easy flow when applied to a fabric. In selection of the base polymer, some of the important properties needed will be thermo-plasticity, mechanical properties, softness/stiffness, compatibility, if polymers are used in combination, good adhesion, abrasion resistance, thermal stability, melting point, UV resistance capability and some others.

Here, we discuss PUR coatings. PUR belong to a group of very durable thermosetting synthetic polymers. The versatility of PUR enables their use for a wide range of applications. They can be coated on textiles, leather or other substrates in solution, as dispersions with solvents or without it, as granules or powder. The degree of softness or hardness can be achieved by using additives.

PUR have good water-proof properties, good adhesion to textiles, durability at low temperatures and it is possible to use it without softeners. It is easy to achieve required levels of viscosity and abrasion resistance. PURs have a very flexible and soft touch. Current manufacturers achieve a very close resemblance and texture to nature.

There are well-known chemical companies that produce a wide range of PUR and PUR systems particularly suitable for coating on textiles. Coaters can also request systems with standard or ones with special properties formulated to meet specific end-use requirements.

Dow Hyperlast in the UK has special grades of polyester polyols designed for reacting with di-isocyanates to produce high-quality PUR such as microcellular elastomers, adhesives, fabric coating systems, foams, surface coatings and high-performance solid thermosetting and thermoplastic elastomers.

8.3.4 Anisotropy of multi-layered coated materials

Anisotropy allows a material to change or accept different properties in different ways as opposed to isotropy which is defined as uniformity in all orientations.

Coated textile products are produced either as artificial leather or laminates, assuming properties of the materials they are made of. Since they are partially made of textiles, that is, in their properties mostly anisotropic, coated materials as a whole are also anisotropic, meaning that the coated material behaves differently in different directions when stressed.

Anisotropy of woven and knitted fabrics is outstanding. This property is reduced by the addition of coatings but not completely eliminated. Thus, the final properties of coated materials do not depend only on the components but also on the direction of load.

8.4 Innovative coatings on technical textiles

The demand for technical fibres and textiles has been increasing from consumer needs to industrial and also to high-tech product needs. Researchers, developers and product designers are constantly coming up with new and innovative high-performance textiles that enhance functional efficiency, comfort and safety in mind. This combined with newer high-tech coatings being available, manufacturers of coated materials have the opportunity to produce newer and advanced coated materials to meet the constant market demands.

Where durability is of prime importance, silicones are the choice for coating of technical textiles. An example is the constant efforts made by Elkem Silicones, who are committed to pushing the limits of what technical textiles has to offer to produce high-performance artificial leathers.

Bluesil coating systems are specialized engineered silicone-coated fabrics. Thus, these coated fabrics can provide the best possible properties for each end application as required. Some of the main properties are based on:

– Tensile strength
– Fire and UV protection
– Resistance to abrasion/flexing
– Long usability
– Provide exceptional comfort
– Environmentally friendly products

Here are some examples provided by Elkem Silicones regarding applications:

Airbags have best and long durability with nylon and polyester textiles as substrates.

Protective fabrics: Bluesil TCS silicone products are exceptionally resistant and withstand the harshest conditions. They provide added insulation, as well as barrier properties against liquids, in particular to medical garments or personal wear.

Sports: specially engineered emulsion formulations for textile coatings make it possible to manufacture lighter, stronger and flexible coated textiles to make sports products that have to meet challenges in the field of sports.

Architecture: silicone-coated fabrics provide high-level mechanical properties, environmental resistance and aesthetic values such as luminosity. Bluesil TCS products also enable to meet the highest safety standards.

According to Elkem Silicones, their experts are available worldwide to provide advice and assistance for manufacturers to improve and achieve exactly what is needed in coating processes to make products of the highest quality.

8.5 Coating methods

Artificial leather is a combination of a substrate (backing fabric) made from natural, synthetic or bio-fibres or a combination of them and a polymeric coating. Thus, the fabric when coated imparts new characteristics to its base fabric which are woven, knitted or from non-woven materials. To manufacture artificial leather, there are many coating methods that can be used to apply a polymeric coating to a textile. In general terms, they can be classified on the basis of equipment used, method of coating, curing and finishing systems. Some of these are as follows:

1. **Coating**: coating polymeric material is in the form of a paste, solution or lattices.
   a. Knife coaters: a simple example is a solid blade 'doctor knife' or air knife over the textile/fabric to be coated. The fabric is under tension and backed by a roller underneath. The knife is set over the fabric and the polymeric material is poured onto the slowly moving fabric and the pre-set gap between the knife and fabric surface will determine the coating thickness. First, a base coat is applied, followed by additional coats as required to achieve the desired final thickness. The last top/skin coat as it emerges out of the final curing oven is then embossed/finished as desired. This is a general pattern for direct coating.

      In indirect or transfer coating, instead of the first coat on fabric, the polymeric material is first coated onto a slowly moving pre-embossed release paper, which will be the top/skin coat. The coated paper is then rolled back to the feed station and the foam coat is applied and a fabric laminated onto the foam surface. The slowly moving paper carrying the laminated fabric will go through a heating/curing system where the foam coat will expand to form a cellular mass sandwiched between the paper and the fabric. The complete embossed coated material is then peeled off at the take-up end. Some processes may have an online trimming arrangement to remove the edges of the uncoated material.
   b. Roller coaters: reverse roll coaters, kiss coaters, gravure coaters, dip coaters and others. These are pre-metered application systems.
   c. Impregnators: material to be coated is dipped in the polymeric fluid and the excess is removed by squeeze rollers or doctor blades.
   d. Spray coaters: the polymeric material is sprayed directly on the web or on to a roll for transfer.
2. **Coating with dry compound (solid powder or film)**
   a. Melt coating
   b. Calendaring
   c. Lamination

The artificial leather industry has seen rapid advances from its original PVC coatings to PUR and then again to silicone-coated fabrics. The advent of graphene has opened up a whole lot of possibilities, and now, graphene-induced polymers are offering newer possibilities where properties are concerned. As the demand for coated materials with more and more special properties is in demand, not only the coaters but the machinery and equipment manufacturers also have to keep up with new designs and effective coating processes to meet final products with large volume and quality at economical costs.

Selection of a coating method depends on several factors such as:
- Final properties and quality of the artificial leather required
- Type of a substrate to be used

- Type of the resin/resins to be used
- End application needs and accuracy of the coating needed
- Economics of a process to be used

8.5.1 General features of fluid coating systems

Different manufacturers of these coating systems may have different features to offer coaters but the basic coating principle is the applying of a fluid polymeric mix on to a slowly moving substrate through a pre-set thickness at the feed end. The coating thickness gap can be set by a thickness gauge manually or by auto setting by sensors, which will be more accurate.

Fabric preparation: Fabrics for coating will generally be in rolls containing long lengths, the longer without joints the better. Joints will create coating problems and a few inches on either side of a joint will be the reject material. There are two common methods of getting a roll of fabric into a smooth coating position. The first one is to thread the roll from the take-up end through the ovens and onto the feed end. This roll is then rolled back under tension which will 'spread' the fabric in both directions – lengthwise and widthwise (warp and weft) – and also give an operator an opportunity to remove any surface flaws like wrinkles, knots, fibres sticking out or other. At the end of this process, the fabric will wound up smoothly and will be ready for coating, while still being connected to the take-up end. Some coaters may have a 'lead cloth' attached to the end of the roll to save some fabric, which will be the entire length of the coating system.

A smaller coater, for example, an entrepreneur, may even carry out this operation on a long table with a fabric roll mounted on one end and rolled out under tension.

Coating station: An important aspect of a coating process could be based on a knife, roll or any other methods of fluid coating. Artificial leather is made in wide widths, for example, 108 in (270 cm) or it can be more. To achieve this final width, the width of the starting fabric will have to be more as it will lose width due to tensioning on the coating system. The extension of the loss will depend on its structure, weight, strength and so on. Therefore, the coating station must be wide enough to accommodate these wide fabrics.

Heating/curing ovens: There are different systems. The type, length and temperature ranges will depend on many factors starting with the type of coating polymer, blowing agents used, evaporation time, thickness of coatings, width of substrate and so on. The ovens should be well insulated and have effective exhaust devices. Using an advanced universal system will be an advantage to a coater to cover a full range of polymers and their processing needs.

In a simple direct coating process, the coating system may need only one oven but for foam production, especially in indirect coating or transfer coating systems, a minimum of two ovens is needed. The design of the heating/curing oven systems is important and some ovens may have zonal heating with gradually increasing temperature zones. In the case of organosols, the drying rate must be carefully controlled to prevent formation of blisters or cracking. The speed or rate of passing through the ovens depends on the coated polymers, their thickness, rate of cure needed on the basis of surface area and the temperatures used must not scorch or weaken the strength of the base fabric. Since in most cases, dyed fabric is used to match the colour of the polymeric coatings, original colour preservation should also be considered.

Winding section: In direct coating, the hot material coming out of the oven will be backed by cool-rollers and if the embossing is to be done at the same time, an infrared heating frame will be brought over the warm coated surface to heat it sufficiently to obtain a good emboss. Here, a word of caution. If the infrared heating device is lowered too close to the coated surface, chances are that it will catch fire due to excess heat build-up. The embossing roller will be water-cooled to prevent too much heat build-up, which will affect the surface embossed grain. Some coating systems will have an edge-trimming device across the width to separate the coated area from the non-coated edges. This operation could be carried out later also during the packing operation.

In indirect or transfer coating, where a silicone-release paper is used, the winding section becomes a little more complicated as it has to accommodate a wind-up of the release paper after the coated fabric has been 'peeled' off. The alignment of the paper wind-up must be accurate and smooth to prevent damage to the paper which is to be reeled back to the coating station and re-used.

8.6 Knife coating

Knife coating systems are one of the most common coating methods. A prepared fabric is fed through a backing roller and the fabric under a 'knife' known as a doctor blade, through a heating oven and secured at the wind-up end. A coating gap is set between the doctor blade and the surface of the fabric. The fabric is set in a forward motion at a very low speed and the prepared coating material is *poured onto the fabric surface at the front of the doctor blade and the slowly moving fabric will carry the polymeric material with the set gap determining the thickness of the coating. An operator standing in front will slowly spread the coating material with a spatula to make sure the coating is fed in a smooth and continuous manner.*

The forward motion of the fabric and the set gap of the knife create a rotary motion of the poured viscous material. This is known as the rolling bank which can be added on, if necessary and will function as a reservoir for the polymeric coating material. To control and prevent this mix from leaking over the edges of the fabric, this 'bank' has two adjustable guard plates at both ends of the fabric known as dams. The fabric must be under tension, both lengthwise and widthwise to maintain a smooth fabric surface to ensure a smooth and even coating. Most coating machines can coat fabrics with wide widths and advanced machines can accommodate up to 4.0 m widths. When using woven fabrics, which will normally have a fairly open weave, though not quite visible, a coater would increase the viscosity of the first/base coat of the polymeric mix to prevent the material seeping through. This is normally achieved by the addition of a thickening agent. The coated fabric will then enter the heating/curing section of the coating line with the required parameters like speed and temperature pre-set to achieve the desired rate of evaporation/semi-cure of the coating.

An important phenomenon worth presenting here is the wetting of the fabric when the polymeric mix is poured onto it. Wetting determines the degree of fabric penetration and also the degree of adhesion between the mix and the fabric surface. When the polymeric mix is poured on to the fabric behind the coating knife, it will spread with the viscosity determining the rate of spreading. The mode of pour, meaning it would be ideal if the material is poured to reach the full width of the fabric at the same time to minimize uncoated 'waste'. In some operations, if the coating material's viscosity is high, an operator could spread the poured material quickly using a spatula. Generally, the wetting property depends on the solvents or liquids used. After a few seconds, evaporation ceases but with slight increase in viscosity and the polymeric mix will settle down for smooth coating.

8.7 Knife coating systems

There are three knife coating systems: (a) knife on air, (b) knife on blanket and (c) knife on roll.

The knife over roll system is the most popular and important technique used widely and adapted for its simplicity and has much higher accuracy over the others. In this system, a suitably designed doctor knife is securely positioned over a high-precision roller. The roll may be a rubber-covered or chromium-plated steel roller. The hardness of the rubber roller may be from 60 to 90 shore depending on the types of fabrics to be coated. The advantage of a rubber-covered roller is that it will accommodate any overlooked fabric flaws like knots, fibres sticking out or other allowing them to pass between the rubber surface and the coating gap because of the 'give' on the rubber surface. However, a possible disadvantage exists in that with constant use the rubber surface may get affected due to material seepage, constant abrasion or swelling or other and may have to be re-surfaced. With rubber rollers, it is possible to have slight variations in coating thicknesses. On the other hand, steel rollers give a firm backing for the fabric and give smoother and more uniform overall coating thickness. However, localized coating problems may occur due to any fabric flaws as there is no 'give' on the steel surface.

The design of the coating knives also plays a key role in the efficient coating of a fabric. Since many parameters like different polymeric mixes, low to high viscosities, different coating thicknesses and weights are involved, it is important to use a knife design which will give the best results.

8.8 Transfer coating

In direct coating, where a polymeric mix is directly applied onto a woven fabric, the penetration of the mix occurs through the open weave of the fabric, increasing adhesion, lowering tear strength and decreasing elongation, resulting in a somewhat 'stiff' coated fabric. Transfer coating overcomes these limitations because no tension is applied on the fabric during coating and the most delicate and stretchable fabrics can be used, for example, knitted fabrics, which have open weaves but very flexible.

Transfer coating, also known as *indirect coating*, consists of a silicone-coated embossed paper onto which a thin first layer or 'skin' is coated and semi-cured. This coated paper is wound back to the coating station and a second layer consisting of a foaming agent is applied onto it. After passing the coating knife, it goes through a laminating unit, where a fabric, usually a knitted fabric, is pressed onto the surface and laminated. This lamination will press the fabric onto the surface of the wet coating and should be done to ensure sufficient adhesion only, and too much pressure applied will tend to 'crush' the foamed layer, thus, preventing rise of the foam to its full height as desired.

As the paper now containing the coated material enters the first heating oven, the foam layer is activated and semi-gelled and when going through the second oven, where the temperature is higher, the foam will rise to its full height controlled by the laminated or by the knitted backing fabric. As it emerges from the oven, there is no embossing operation as the embossed release paper has already provided it. The coated material is taken off the release paper and both wound on separate winding stations. This material will have uncoated 'edges', which has to be trimmed, either on line or at a later stage, for example, during inspection and packing. If printing or other finishes are desired they can be carried out as separate operations.

Some advanced coating systems may have two coating stations on line and this will enable foamed artificial leather to be processed in one operation. Release papers come in very large rolls, heavy and wide. They come in different qualities, generally, based on their paper thickness/weight. Naturally, the thicker and heavier ones may last longer but all qualities are expensive. Therefore, careful handling on the coating line is required to ensure one roll can be used at least 5 to 6 times. Since these release papers are reeled in forward and backward motions during a coating process, some problems to guard against are scorching and tears.

Release papers are manufactured to a uniform thickness and coated with a thin layer of silicone for easy peel off of coated layers. A main feature of a surface of a release paper should give a good emboss pattern and also release the final coated and fabric laminated product off the paper without any damage or distortion of the emboss. Estimated number of times a roll of release paper that can be re-used is generalized as 8 to 10 times but from experience, the author suggests 5 to 6 times.

A coating station in a transfer coating system will typically contain a knife over rubber roller, acting as a backing roller. A rubber roller has an advantage over a steel chromium-plated roller in that it is less likely to damage the release paper. The laminator rollers are steel or rubber coated and the gap – *the nip* – can be pneumatically or electronically set to suit the production of different man-ufactures. Similarly, the coating knife angle, the coating 'gap', can be set and op-erated automatically.

The exhaust systems in each oven must be very effective, especially when sol-vents are used to prevent possibilities of fire and general air quality of the plant. Modern ovens are well designed for most polymers for coating and it may be ad-vantageous to invest in universal ovens. The machinery and equipment industry for making artificial leather being well advanced, coaters can have a choice of in-vesting in standard coating lines or custom-made processing lines.

8.9 Hot melt coating

This is a process by which a hot polymeric mix is applied by extrusion onto a fabric through a wide T-die.

8.10 Gravure coating

Gravure coating systems use a pre-calculated amount of a polymeric mix as a coat-ing on a substrate.

8.11 Curtain coating

This is a coating system which uses an uninterrupted vertical flow of a polymeric mix to fall onto a substrate which is under motion. The thickness of a coating is controlled by an adjustable die gap of the holding tank and the speed of the moving substrate.

8.12 Dry powder coating

Coating method is used for polymer powders that can be fused together for special applications.

8.13 Dip coating

In the polymer industry, dip coating can mean a very common and useful process where metals, plastics or other things are 'dipped' in fluidized polymer solutions for protective or enhancing aesthetic values. However, coating of textiles, whether they are made of natural or synthetic fibres, also forms an important branch of the coating industry. This is a useful method of coating substrates on both sides.

Dip coating is a process based on a substrate immersed in a tank of the polymeric coating material for a certain period of time, known as the dwell time. As the coated textile emerges out of the tank, the excess material is then squeezed out by passing through a set of nip rolls or a set of flexible doctor blades, pre-calibrated to achieve a net pickup of the coating material. There could be various arrangements for a dip process but the basic principle is the same. This process is ideal for making artificial leather coated on both sides of a substrate.

8.14 Reverse-roll coating

This method can be used for a wide range of coatings of different viscosities and coating thicknesses with high coating accuracy. Reverse-roll coating systems apply a pre-metered coating of uniform thickness, regardless of any variations in substrate thickness and are, therefore, known as contour coaters also with the coating independent of substrate surface. There are two basic systems: one being the 'three-roll-nip' and the other the 'pan-fed' system.

8.15 Slot die coating

Slot die coating is a coating technique used for the application of solutions, slurry or extruded thin films directly onto typically flat substrates such as glass, fabrics, metals, paper and plastic films or sheets.

8.16 Self-healing polymeric coatings

Products used for industrial and consumer applications are usually coated to improve appearances and protect them from usage and from weathering effects like ultraviolet light, rain, mechanical and chemical effects.

NEI Corporation reports a development in self-healing nanocomposite polymeric-based coatings. Polymer coatings impart three important functions to an underlying substrate: the aesthetic function which gives a substrate a good quality and feel and appearance; the protective function, which gives a substrate good resistance and prevents mechanical and chemical damage; and also durability. Constant use will generate surface scratches and micro-cracks, which eventually leads to macroscopic damage and the coating losing its aesthetic and protective functions.

NEI's patented coatings are unique in that to increase longevity, when heat is applied they will self-repair or self-heal minor surface scratches and relatively deep cracks. NEI's patented coating technology has been used for repeatable self-healing on solvent-borne PUR coating systems. The coating industry is moving towards solvent and volatile organic compound–free waterborne and ultraviolet curable coatings.

Author's note: As identified earlier in this presentation, there are other different coating systems and it is not possible to present all. However, readers should be aware that most coating systems are based on the principle of direct and indirect coating techniques which are the most popular processing methods used by manufacturers of artificial leather. Due to constant research and development, existing coating systems may see modifications for more efficient processing.

Bibliography

[1] Tex Tech Industries. publication "All You Need to Know about Polymer Coating Systems." https://textechindustries.com, 2015 Home Blog.

[2] GARPH Publication-Academia.edu. "Nonwoven for Artificial Leather." vol.2/No.2/ February 2013, www.garph.co.uk

[3] Stana Kovadevic, Darko Ujevic and Srijezana Brnada. article "Coated Textile Materials." University of Zagreb – Croatia – www.intechopen.com

[4] Madara Jaskova. article "Coating Methods." Riga Technical University – Textile Technology and Design – Riga Latvia, 2008.

[5] Elkem Silicones: Publication. "Technical Textiles – innovative coating for technical textiles." https://www.elkem.com

[6] Chris Defonseka. Book Title "Processing of Polymers." De Gruyter: pages 90–91, www. degruyter.com 2020

[7] Publication: "Gravure Coating Equipment". New Era Converting Machinery Inc. https://www. neweraconverting.com.

[8] Coating Tech-Slot Dies. publication: "Curtain Coating Technology." https://www.slotdies. com/curtain-coating-technology.

[9] Dr. Peter M. Schweizer. "Curtain Coating Technology." https://pffc-online.com/mag/paper_ curtain_coating_technology, March 2000.

[10] Arville Publication. "Textile Coating Process." https://www.arville.com/technical-textiles/ coatings/processes. May 2018

[11] Publication: "Reverse Roll Coaters." New Era Converting Machinery Inc. https://www.newera converting.com

[12] Dr. Adrian Hill. article "Rheology for Coatings." PCI Paint & Coating Industry March 2006.

Chapter 9
A small manufacturing plant for artificial leather for an entrepreneur

9.1 The concept

An entrepreneurial venture to make synthetic artificial leather will be both exciting and rewarding. As an entrepreneur, the manufacture of a wider range of coated products, although not strictly coming under the category *artificial leather* is possible using the same machinery and equipment. For large volume manufacturers, they may not have the production time or it may not be lucrative enough to produce these types of products. However, for an entrepreneur, these polymeric coated products will be an additional profitable income. At least a basic business/technical background will be helpful to start with. The market potential is very important and naturally one would carry out a detailed market survey/analysis and select a few basic products for manufacture initially.

If resources are limited, one would look for bank assistance also, and in this case, it is better to look for a package where the cash content should be minimal/manageable but with maximum facilities where they would have to be paid back, if and when used only. First step would be to produce quality materials readily acceptable by the marketplace. You may have quality products but customers may not exactly come running to your doorstep to purchase. You must also have an effective marketing strategy.

Manufacturing of artificial leather is considered as a big time operation, the market for these products being broadly divided into two types: solid coated synthetic leather and foamed coated synthetic leather. A third may be considered coated/embossed synthetic substrates and coated bio-fibre products, although the market volumes for the third may be much smaller.

As we know, the three main polymeric coatings currently in big demand are for coated substrates with polyvinyl chloride (PVC), polyurethane (PU) and silicones. It is recommended that as an entrepreneur, the targeted production range will be confined to solid PVC coated leather and also the products mentioned in the third category. The volume demand for these products at any time should be sufficient for an entrepreneurial venture to be viable and very profitable. PU and silicones are not considered in this presentation as they will require advanced manufacturing technology and more sophisticated coating machinery and equipment.

This presentation is designed on the assumption that an entrepreneur's initial investment will be sufficient for a viable venture with reserve resources for expansion, and the entrepreneur has a sound knowledge of polymeric coating technology or will have access to at least one knowledgeable person. The products to be manufactured

https://doi.org/10.1515/9783110716542-009

will be in accordance with British Standard Specifications or other to be selected by the entrepreneur in accordance with the markets targeted.

9.2 Recommendations for infrastructure basics for a production plant

For the benefit of readers, the author presents here an ideal design for an entrepreneur to manufacture artificial leather coated with PVC on a small scale, based on an actual plant operation set up by him. This plant can produce a wide range of synthetic leathers plus other coated or embossed products for which a great market demand exists.

Only the important sectors for manufacturing are discussed below and the rest is left to the imagination and initiative of an entrepreneur:
– Range of products
– Market share and marketing strategy
– Factory space
– Machinery and equipment
– Plant layout
– Engineering services
– Personnel
– Technology
– Raw materials
– Formulations
– Production methods
– Production hazards
– Safety wear

9.2.1 Range of products

A wide range of products is possible but for an entrepreneurial venture perhaps a limited good quality product range would be the best for starters. Primary targets for products would be for upholstery, footwear, handbags, fashion wear, safety wear and some consumer items like shower curtains, coated paper for packaging and coated bio-fibre cloths for handicrafts or other. If production times permit, perhaps coated materials for tarpaulins or others can also be included. Since it may not be possible to produce all types at the beginning, an entrepreneur will no doubt select products that have the highest market potentials and also the biggest contribution margins for good profitability.

This operation will use standard to high-end cotton fabrics, depending on the end applications and the latter qualities will be dyed to match the colour of the coatings which will greatly enhance the aesthetic values of the products and thus the quality also. These artificial leathers can be produced with surface finishes, plain or embossed in matt, glossy, lacquered or transparent coated finishes. Packaging will depend on current required standards and all packing labels should include an ID for the standards they are being made to.

9.2.2 Market share and marketing strategy

Before starting this venture, during the preparation of a feasibility analysis an entrepreneur would naturally be well aware of the markets and this potential market share to be targeted. A breakeven analysis (sales needed to cover total costs) will also be helpful in determining the minimum amount of sales needed. Again, the contribution margin for each product towards meeting costs will also be helpful in determining which products are feasible. Here, it is important to note that some products may have very large sales volumes with low contribution margins, while others may have smaller sales volumes but with high contribution margins.

An entrepreneur will have several choices of marketing tools like personal contact for direct sales, flyers, advertising, agents and distributors. In consideration for effective cash flow, the best approaches would be direct sales and custom orders but still some kind of credit may have to be extended. For an entrepreneurial venture, a starting market share of around 5–10% of a localized market share may be practical.

Since, in all probability, the marketing budget will be limited at the beginning, low-cost but well-designed flyers and swatches (sets of small samples) will be very useful and effective for distribution among prospective customers. A certification of quality by a recognized organization will go a long way in successful marketing. It is to be expected that an entrepreneurial venture or any manufacturing project for that matter will initially have some 'teething' problems but this can be overcome to a great extent by an effective communication/feedback system with the customers. Generally, there should not be any shortcomings; yet, some of the problems could be colour fading (especially reds), dimensional irregularities, surface blemishes and coating surface brittleness after some time or some other.

Examples of contribution margin and breakeven point:

(a) Contribution margin – amount of money a product contributes towards meeting overheads.

C. M. = unit sale price – unit variable cost

(b) Breakeven point – total sales needed to meet total expenses

Using formula, BEP = TFC ÷ (USP – VCUP)

where   BEP is breakeven point (no. of units),

      TFC is total fixed costs,

      USP is unit selling price and

      VCUP is variable cost per unit product.

Illustration:

A polymeric coater makes artificial leather. On an average, each metre retails at $5.00. It costs $2.00 to make it and the fixed cost for the period is $750.00. What is the BEP in metres and also the sales volume for that period?

Applying formula, X units = FC ÷ (SP − VC)

$$= 750 \div (5{-}2) = 250 \text{ m}$$
$$\text{Sales volume} = \text{BEP} \times 5 = \$1{,}250/\text{-}$$

9.2.3 Factory location and space

To minimize initial costs, an entrepreneur may opt for renting an existing building having a space of about 1,600 sq. An ideal building would be an industrial unit with the necessary local approvals for an industrial operation having three-phase electrical power, water supply, a front and rear entrances, a small office and so on. This building should have adequate lighting and also must have easy access to main roads.

The factory space must be so designed to accommodate machinery and equipment, a mini-lab (optional), a small workshop (optional), raw material storage, finished goods storage, receiving and shipping facilities, some facilities for workers, fire extinguishers at strategic places and an effective exhaust system.

9.2.4 Machinery and equipment

The basic machinery and equipment for direct coating would be as follows:

Coating machine will consist of a material feeding end in front, a heating oven, take-up end with embossing unit and a 'doctor-knife' coating system. If foamed PVC coatings are to be produced, a coater may need a two-oven heating system. For an entrepreneurial venture, a basic manually operated but complete coating machine would suffice.

(a) Material feeding end: this will consist of two main systems. The first is a fabric feed arrangement, where a large roll of fabric will be mounted on a horizontal bar, the fabric treaded through the oven and secured at the take-up end for free movement forwards and backwards.

(b) Heating oven: infrared heating is the easiest with good heat controls to a maximum 250 °C and an exhaust with chimney on top of oven which will be well insulated.

(c) Take-up end: this will have a system where the coated fabric can be rolled up and also fed back to the coating end. As different speeds will be used, both ends will have effective speed control systems to ensure smooth movement.

(d) Embossing unit: this consists of an embossing roller, a backing roller and an infrared heating system. The coated fabric will pass between the backing roller and the embossing roller which can be raised up and down. The heating unit also can be moved up and down to get just sufficient heating on the coated surface to obtain a good emboss.

(e) Doctor-knife system: this is mounted at the feed end of the coating machine will consist of a steel blade of a particular design which can be adjusted to get different angles on top surface of the fabric with a backing roller. This 'doctor-knife' can also be adjusted vertically up and down, which will determine the thicknesses of the different coatings.

Embossing rollers: these are very expensive and every care must be taken to protect them even in storage. An entrepreneur may opt for at least three embossing rollers to start with: certainly one 'plain' roller, a must, plus one for upholstery and another for general-purpose markets. The choice of embossed surface patterns will depend on the market demand. The operation can add more embossing rollers as the project progresses.

Triple roll mill: a standard triple roll mill is suitable for mixing chemicals. This must have emergency stop and preferably a sensor arrangement to work a trip-switch to stop machine immediately in case of an operator leaning too far 'into' the running rollers. This will prevent accidents like an operator's hand or fingers getting caught between the rollers. Operators must wear safety aprons and loose clothing is not allowed.

Planetary mixer: this is used for preparing the coating mixtures, and should have speed adjustments and removable blade for cleaning and emergency stop.

Equipment: since this operation is more or less a manual operation, sophisticated equipment is not required but still the basics are needed as follows:
- Mini-lab/workshop needs/tools
- A few trolleys for transport of material/fabrics/coated fabrics
- A few large plastic or preferably steel containers
- Weighing machines
- Thickness gauges
- Safety wear for operators
- One or two measuring tables for coated fabrics packing
- Office and worker's restroom needs
- Fire extinguishers
- Stand for embossing rollers
- Racks for storing coated fabric rolls
- Any other needs

A preventive maintenance system for all machinery and equipment, however small it is, will be helpful in minimizing downtime. Log books maintained in strategic places on the production floor will greatly help in improving operational efficiency.

9.2.5 Plant layout

The author prefers to leave this exciting and challenging function to the entrepreneur. The overall layout of the plant should ensure 'free-flow' production for maximum efficiency. Provision of at least some basic facilities for the personnel should be considered. Plant security is another important factor.

9.2.6 Engineering services

The three main supplies needed are electrical power, water and air (optional). In order to enhance operational efficiency of the plant, an entrepreneur may want to have the personnel trained in dealing with emergency situations like accidents, fire or chemical spills and these could be done by experienced outside personnel at next to no cost. For example, for dealing with fire the local fire department would probably oblige with training and demonstrations at no cost. However, these areas should be considered as ongoing projects and priority given to producing quality artificial products and successful marketing.

Since electrical power supply is a key element in this operation and for future expansion, a good understanding of it would be useful as shown below, which will help to plan immediate power requirements and also for expansion, if any, for stage 2 and maybe stage 3.

9.2.6.1 Calculating electrical power

Electrical power is one of the key, if not the most important need of any manufacturing operation. An entrepreneur will do well to understand the basics and be able to plan for future requirements also, if expansion of the plant is envisaged. The following guidelines will help in calculating electrical power:

All machinery and equipment operated by electrical power will carry the manufacturer's specification sheets which will include the horse power (HP) and working voltage. For example, a motor will be described as a single phase 1-HP motor or three-phase 5-HP motor. Electrical power consumption costs are based on a basic unit, kWh (kilowatt hour), the quantity of electrical power consumed in 1 h. Voltages may vary from 110/120 to 230/240 V for single phase and 380/480 V for three phase depending on the location. When installing or requesting a power source, it is recommended that an additional 40–50% be added to accommodate starting all motors at the same time. When planning power requirements, it is always prudent to give consideration for possible future expansion programmes.

Some of the basic terms used in electrical power are:
- Amperes (I)
- Current in amperes (I)

- Voltage (V)
- Efficiency (Eff)
- Power factor (pf)

As we know, 1 HP = 746 W. If a motor is rated at a true HP, then it will deliver 746 W of mechanical power. However, single-phase motors are never 100% efficient in converting electrical energy to mechanical energy, so the amount of electrical power consumed by a motor is considerably higher than the mechanical power delivered. Due to losses from heat and friction, a typical single-phase motor will work only at around 60–70% efficiency. For most power tools with induction motors, an efficiency factor of 60% is more realistic. Hence, a genuine HP motor will require ≥ 1,250 W of electrical power to deliver its rated power. This means, that a 1 HP motor will need ≥ 10 amp. Current at 125 V or 5+ amp at 250 V to realistically deliver a true 1 HP form the same motor. This is a good rule of thumb.

The power factor of an alternating current (AC) electrical power system is defined as the ratio of the active (true) power to the apparent power where (a) true power is measured in watts and the power drawn by the electrical resistant of a system doing useful work and (b) apparent power measured in volt-amperes and the voltage on the AC system multiplied by all current that flows in. Table 9.1 shows typical values for motors with different HP.

Table 9.1: Typical values for motors.

HP	Eff	pf at full load	Quantity	kW (SP)	Amp (SP)	kW (3P)	Amp (3P)
2	79%	0.84	2	3.7	19	2.1	11
5	84%	0.84	1	4.4	44	2.5	26
15	86%	0.86	1	13.0	126	7.5	73

9.2.6.2 Example as a guideline
A small manufacturing plant has
- One 2 HP motor
- One 5 HP motor
- 20–100 W lights

Then, power consumption = $(2 \times 746) + (5 \times 746) + (20 \times 100)$

$\qquad\qquad\qquad = 1{,}492 + 3{,}730 + 2{,}000$ W

$\qquad\qquad\qquad = 7{,}222$ W $\div 1{,}000 = 7.222$ kW

Therefore, power consumption of all work for 8 h:

$\qquad\qquad\qquad = 7.222 \times 8 = 57.776$ kWh

If all motors start at the same time, add 50% additional load in watts.

Now, W = volts × amp × power factor

7222 = 440 × amp × 0.8

Therefore amp = 20.51 per phase

A three-phase power of 440 V/50 Hz wiring system to suit 30 amp per phase and a three-phase power of 440 V circuit-breaker trip switch with individual start/stop switches at point of power entry should be provided as a safety measure. If plant is located in an area where frequent power stoppages occur, a stand-by suitable power generator will minimize production downtime. Power outlets at strategic locations are needed.

If air supply is required, compressors can be used either installed as a combined unit or as a single portable unit or stationary with a few outlets for use at different points.

9.2.7 Personnel

With the entrepreneur as the manager and in overall incharge of operations, it is recommended that a floor supervisor and a team of eight operators would suffice. It is prudent to recruit some people with technical backgrounds and preferably with experience in 'plastics' manufacturing, although the latter is not that important as they can be easily trained.

Ideally, the supervisor will have some experience in a plastics manufacturing environment and safety standards. As operations go forwards, all operators should be trained to work in any section of the plant. As an entrepreneurial venture and maintaining costs at minimum levels are important, it may not be possible to have extra operators to counter absenteeism.

As the manufacturing operations take place, after some time, it may be helpful for the entrepreneur to request one or two of his/her close industrial associates or others to see the operations and make helpful suggestions to improve it. As a general rule, it is an independent outsider who could see the 'flaws' and suggest corrective action for cost-cutting and improve efficiency. Good examples are better production 'flow' for cutting costs and colour matching of pigments and dyes for good 'matches' and durability factor, especially for custom orders.

9.2.8 Technology

Probably, technology is the most important factor for the success of the venture. Here, the entrepreneur will play the key role and provide all necessary technology for the project and must have a thorough knowledge of all aspects of manufacturing artificial leather. It would be helpful, if samples of competitor's product data like quality, colour ranges, prices, packaging and others are studied before commencement of operations.

No doubt, material suppliers will give their support and provide data and also perhaps free laboratory facilities. However, actual manufacturing is quite different and needs special skills to produce good quality products.

An entrepreneur will have a thorough knowledge of chemical components to be used formulating, colour matching, fabrics, operation of machinery and equipment and so on or will at least recruit someone who is competent but it will mean an additional cost which could be expensive.

A better alternative for an entrepreneur is to discuss/train with the suppliers of materials, machinery and equipment, which in most cases is cost-free. After all, you are purchasing their products.

9.2.9 Raw materials

For preparing PVC coatings on fabrics to make artificial leather, there are several chemical components required. There are several suppliers made up of manufacturers, agents, distributors and also companies who will supply custom-made PVC mixtures for any type of coatings. If an entrepreneur does not have the necessary knowledge to purchase individual components and formulate and mix them on the production floor, these custom-made mixtures can be purchased in ready-to-coat modes.

Better system is to be able to formulate on the production floor, including colour matching but initially it may be a somewhat difficult task as in addition to the required technology, valuable capital may be tied down. To enhance project liquidity all materials needed, including fabrics can be purchased on *just-in-time* (JIT) basis but here if supplies are held up and not delivered on time, valuable production time will be lost and also may affect customer relations.

The following are the basic components required:
- Fabrics
- PVC copolymers
- Graft polymers
- Plasticizers
- Fillers
- Pigments/dyes
- Lubricants
- Heat stabilizers
- Thickening agents
- Blowing agents
- Additives

A coater can make different types of products by combining these components in different proportions. The polymer industry has been developing very fast over the years and chemical giants like DOW, BASF, Bayer, BP Chemicals and Monsanto are constantly offering the markets newer and exciting products to meet challenges demanded by the various industries, more so, by the artificial leather end users, especially now that environmental concerns are also a factor. An entrepreneur will do well to contact these companies or their agents for advice, suggestions, formulations and so on, which will give an entrepreneurial venture a tremendous boost and confidence.

9.2.10 Formulations

One of the key functions for manufacturing synthetic artificial leather. There are two basic formulating systems for an entrepreneurial venture. If the targeted end applications are for upholstery and other general-purpose applications, one basic formula would suffice and the slightly different qualities required could be varied by type of fabric, colour, overall coated thickness and embossed patterns.

However, for more sophisticated materials for high-end applications, for example, footwear, automobile upholstery or other special formulations may be required, including foamed PVC. For embossed synthetic films, packaging and protective wear, they are straightforward. For production of some psychedelic PVC coated material, one could save the different coloured 'left-over' PVC mixtures from the main production runs and use them.

Table 9.2 shows the four basic formulations required, which can be used as guidelines.

Table 9.2: Guidelines for formulating PVC coatings.

Components	Base coat	Middle coats	Skin coat	Transparent coat
PVC polymer	85.00 pbw	85.00 pbw	100.00 pbw	100.00 pbw
Filler polymer/filler	15.00	15.00	–	–
Plasticizer	60.00	70.00	75.00	80.00
Lubricant	00.30	00.20	00.20	00.30
Heat stabilizer	00.10	00.10	00.10	00.20
Thickening agent	05.00	–	–	–
Colour	To suit	To suit	To suit	No colour
Additives	00.20	00.20	00.20	00.30

Note: If foamed PVC is desired, a small amount of a compatible blowing agent should be incorporated into the middle coats. Components are indicated as parts by weight (pbw).

9.2.11 Production methods

If chemical components are to be formulated on the production floor, the filler/fillers and pigments or dyes can be mixed thoroughly on the triple roll mill until a very fine homogeneous coloured mass is obtained. This can be repeated for many colours for standard productions, and they can be stored in large containers to be weighed and used for each coating mixture.

The PVC and the plasticizer according to the formulation being used is put into the planetary mixer and mixed well. A weighed amount of the filler/colour batch is then put into the PVC mix and then mixed well until a homogenous mix is obtained. The other components are then added with the desired additives being added last and mixed well. A separate similar batch is prepared but with a thickening agent and this mix will only be used for base coats only. If foamed leather is to be produced, a separate batch with a blowing agent will be prepared to be used as a middle coat. For high-quality products, the top coat or skin coats, batches prepared will not contain any fillers.

The roll of fabric will be mounted at the feed/coating end of the machine and the 'lead-'cloth will be passed through the oven and secured at the take-up end. The 'doctor-knife' system will then be lowered onto the fabric surface and the two 'limits' on either side set just inside the fabric borders. The gap between the doctor-knife and the fabric is now set, which can be set either electronically or manually with 'feeler gauges'.

The base PVC coat containing the thickening agent will then be poured onto the 'trough' space created by the doctor-knife and a supporting roller used to hold the fabric in a horizontal position. The heaters in the oven are switched onto ensure only a semi-cure as the coated fabric passes through and is rolled up at the take-up station. The heaters are switched off and this cloth is then rolled back to the feed end for the next coat. The coating gap is re-set and the second coat is applied and semi-cured again. The number of coats will depend on the final thickness desired.

The top and final coat will be specially formulated with colour and passed through the oven at very slow speed and high temperature to fully cure the PVC and as it emerges out of the oven, the embossing roller system is gently lowered onto the coated surface along with another heating source to soften the top coat and obtain a good emboss. If additional transparent protective coats or lacquering are to be applied, the process will be similar but without the embossing rollers being used.

If foamed artificial leather is to be produced using the direct coating method, the base coat will have the thickening agent, the middle coat will contain the blowing agent and the surface coat will in all probability be standard. Here, caution must be exercised that during the embossing process, the warm/hot foam is not compressed too much which will result in loss of overall thickness. Alternatively, a coater may apply a slightly thicker foam coat to compensate loss of thickness due to embossing. For foamed PVC artificial leather production, the indirect coating system is the better choice.

9.2.12 Quality control system

As in any manufacturing operation, a quality control system is required here, however, simple it will be. Some of the important parameters to check will be:
– Weight of fabric according to specifications
– Weight of coated fabric
– Effective width of coated fabric
– Coated thickness
– Variations if any of coating across width/length
– Coated surface flaws
– Colour fastness of coated surface
– Length of coated fabric roll
– Packaging labels
– Any other

Quality control should always be done by an independent person like a QC inspector. To keep costs down at the beginning, a staff member could do this, for example, a shipping clerk or other. A simple start for a QC operation would be a colour code system such as
– Green – passed/acceptable
– Yellow – on hold
– Red – reject

In this production operation, all the packaging labels of acceptable products should carry a green label (ideally a sticker) before being sent to storage/shipping department. For goods on hold, they could be held in a large bin or other having a large yellow sticker or painted yellow, until the supervisor checks them. As for the red bin, further inspection may be carried out to see whether some lengths of the finished coated material which are acceptable can be taken out, although they may be short lengths. Even these will have a market with small end-users.

Although the products will be made to conform to some local or preferably an international standard, the floor QC is restricted to conformity to the organization's pre-determined standards only, as there is no way to compare them to the international standard until samples are tested by a recognized outside source. Some of the basic tests will include coated weight, overall coated thickness, tear strength, colour fastness and ageing.

9.2.13 Production hazards

Since the plant will be dealing with chemicals, heating and machinery, certain inherent hazards have to be expected. Starting with chemicals, proper storage, good overall ventilation for the plant and effective exhaust system for the heating oven/ovens and dealing with possible chemical spills would be essential. All operators must be protected from the chemicals being used, although they are not as hazardous as PU chemicals.

The heating ovens and the embossing station will be using high heat, and precautions must be taken to control and ensure the heating temperatures are within limits. At the embossing station, it is likely that different levels of heat will be used (e.g. thin/thick coatings) to obtain good embossed patterns. If the heating device is lowered too low (close to the coated fabric), it will catch fire. A good exhaust is needed to take away the toxic fumes generated by the heating of PVC.

As for machinery, all moving parts must be properly covered and the operators well trained in the operation of all machinery and equipment. The triple roll mill needs special attention and protective devices to prevent an operator's hand or a loose sleeve or other getting caught between the rollers. The planetary mixer will work from slow to high speed and for sampling or adding additives, the machine must be stopped completely. Never add or try to take out any part of the PVC mix while the machine is in motion.

It will be prudent to have at least basic first aid assistance as minor burns or others are to be expected. It will be helpful if at least the supervisor is familiar with the material safety data sheets.

9.2.14 Safety wear

Safety of all personnel is important on a production floor. It is fairly common for some operators to be negligent and not wear safety wear given to them. It is the responsibility of the floor supervisor to see that all operators are wearing proper safety wear.

Some of the safety wear suitable for this production operation are safety shoes, goggles, masks, gloves and aprons.

9.2.15 Safety factors

At the beginning of production operations, an entrepreneur may not be able to implement everything to ensure smooth and efficient production but should be at least aware of the standard needs and implement them as the project progresses.

Starts with emergency exits and also availability of suitable fire extinguishers at strategic locations. Adequate ventilation is important and periodic air quality checks (optional) by an outside agency would help. The raw materials must be carefully stored at temperatures as recommended by technical data provided by suppliers and for PVC raw materials room temperatures are satisfactory. The shelf life of these materials must also be a factor.

Chemical spills are to be expected and the project must be able to cope with them. Provision should be made for general material waste disposal. Posters at strategic locations within the factory like no smoking signs and others would also help. Polymeric coatings, especially PVC and PU, will give off toxic gases and even though the exhaust systems on the machines will work well, for overall long time safety, it is recommended that periodic tests of air quality inside the factory be carried out by an external professional source.

Bibliography

[1]   Defonseka, Chris. "Practical Guide to Flexible Polyurethane Foams." Smithers Rapra. 83–85, 2013.

Chapter 10
Safety factors and production efficiency

10.1 Safety factors

Safety factors are important in any activity and so much so in any industrial manufacturing operation. In a manufacturing plant for artificial leather, safety concerns may be slightly different to, say, a plastics injection moulding or extrusion plant. First of all, the floor space calls for length to accommodate a complete coating and finishing line. Some may opt for coating and embossing/finishing as separate operations, in which case, additional surface heating/warming equipment will be needed.

A typical production floor will follow the standard essentials like good ventilation, equipment for emergency fire hazards and effective exhaust systems. A fully trained crew for fire and chemical spills would be an advantage. For fire drill the best source for training would be the local fire department, while for chemical handling, spills and others an organization could follow the material safety data sheets (MSDS) standards.

On the basis that polyvinyl chloride (PVC) and polyurethane (PUR) are the bases for polymeric coatings, while PVC raw materials will not pose any problems in storage, the PUR will need some safety storage and handling due to its toxicity and flammability. PVC comes in 25 kg bags or in larger packs, and the other additives, liquids or powders will not pose any problems in storage. However, during curing of the PVC coatings under heat, toxic gas will be released and adequate exhaust systems are needed. The PUR raw materials are liquids and generally come in steel drums with two 'bungs' on the top, one large and one smaller. Before drawing out the contents, it is important to let off the pressure built up inside by slowly opening the smaller bung. After drawing the polyol and the isocyanate from the drums, it is important to close both bungs tightly to prevent any moisture getting in. If a manufacturer opts for fully mixed/blended PUR systems, the hazards would be less but the range of manufacture will be limited to those specifications only.

PUR chemicals are termed 'hazardous', and it is important to follow established basic guidelines in the proper handling of these chemicals, procedures for disposal of these post-production chemical wastes and methods for containment and clean-up of possible spills. The two main components – *polyol and isocyanate* – should be handled carefully. Typically, polyols are mild irritants to the eyes and skin and exhibit very low oral toxicity. Any contact with eyes or skin should be washed off immediately. Isocyanates are more hazardous and MSDS will provide information on how to handle them. Therefore, the need for eye-wash stations and one or two shower stations are important.

https://doi.org/10.1515/9783110716542-010

During the installation of machinery and equipment, all moving parts must be covered and each machine must have an emergency switch-off button and an over-all emergency button to stop all machinery in case of a fire or accident. All machinery must be well insulated and electrically safe. Since triple roll mills will be used, it is imperative that a safety device be provided to prevent operators leaning forward too much with the possibility of a hand getting caught between the fast rolling cylinders. A bar mounted across the mill on the side of the operator and electrically/sensor operated would stop the mill if an operator leans forward too much and touches the bar. As an additional precaution, operator could wear skin-tight gloves. Caution is also needed when operating planetary mixers for blending and mixing purposes as the high torque of the rotating blades can injure an operator. It is imperative that each time an additive or other is added, the machine should be shut off and re-started.

During transport of raw materials from the storage area, especially the PUR chemicals, static electricity should be avoided as sparks would be harmful to the flammable liquids. If pre-mixed chemical systems are used, the danger may be somewhat less but still caution is needed.

In complete coating lines, in both direct and indirect coating, the warm PVC or PUR-coated surface that emerges out of the final curing oven has to be heated to a higher degree by means of an infrared or other heating system before it goes through the embossing roller. However, if the heating device is lowered too close to the surface it will catch fire. Therefore, a limit switch or something simpler should solve this problem.

The wearing of safety equipment like goggles, respirators, gloves and aprons, especially for the preparation of the coating plastisols like milling, blending and mixing and coating line operators should be mandatory since they would be dealing with hazardous chemicals and also prevent harm from toxic gases emanating from the curing ovens, although these will have their own exhaust systems.

The production floor should display posters for safety practices, no smoking signs and any others to make everyone aware that good safety practices are essential. For the benefit of the production floor operators and supervisors, management should carry out periodic air quality analyses to make sure that the air quality is within the acceptable industrial standards. This is best carried out by an outside professional agency who could also see any shortfalls in safety factors, which management would generally not observe.

10.2 Production efficiency

In general terms, *productivity* can be described as the *input/output ratio* and there are different ways of evaluation. Probably, the most common one is to evaluate in monetary terms. Good results for productivity depend on many factors and in this case, in an industrial operation, *efficiency* will play a key role. Efficiency will depend on all departments of operations where some will yield high efficiency and others less. However, on an overall basis, if a manufacturing operation can achieve around 80% efficiency, it can be deemed as acceptable.

Some of the drawbacks for under achievement are lack of skills, frequent machinery breakdowns, time loss due to power failures, unavailability of raw materials, accidents, chemical spills, lack of coordination between departments, poor maintenance of machinery and equipment, lack of monitoring and corrective action among others.

Some of the areas for boosting efficiency on a production floor are:
- Understanding of an organization's objectives and quality standard targets
- Motivation posters at strategic places for safety, preventing accidents, awareness of hazards, fire warnings and no smoking signs
- Machinery: proper installation, operator training, maintenance, start-up and shutdown procedures
- Quality assurance: quality control system, documentation, analysis and corrective action
- Process engineering: time and motion studies, analysis and constant process improvement
- Lean manufacturing practice: eliminating unnecessary work, cost cutting and raising profits levels
- Tools: provide correct tools and proper handling, and safety factors
- Workforce: good morale, motivation, good attendance, promote teamwork and safety
- Achieving pre-set targets: productivity and minimizing waste/rejects
- Raw materials: proper storage, shelf-life and proper formulating
- Finished goods: proper packaging and storage system for easy identification
- Good management: communication, workforce motivation and incentives
- Efficient production flow, minimize downtime and proper waste disposal
- Job rotation for operators where applicable
- Correct selection of personnel, counter absenteeism and attendance bonus
- Constant monitoring by plant manager, supervisors and lead-hands
- Proper action: spill management, accidents and fire

10.2.1 Preventive maintenance

It is imperative that a production floor should have a well-planned preventive maintenance system and implemented effectively. At all times, adequate spares should be available. The availability of a small in-house workshop with at least a mechanic and electrician will help. Log books in addition to the production ones at each machine are a must.

Preventive maintenance means that the maintenance crew will inspect, check oil levels, safety of moving parts and so on, as a continued action on a daily basis. One of the major problems for manufacture of artificial leather is the back-up rollers used at the embossing station. The slowly moving coated cloth as it comes out of the final curing oven will pass between an embossing roller and a back-up roller in a synchronized motion. To obtain a good emboss/pattern, the embossing roller will be pressed hard on the coated surface and tremendous heat will be transferred onto the surface of the back-up roller. Generally, hard rubber or synthetic rubber rollers are used because a certain amount of 'give' is needed at the point of pressing. Due to this after some time, the rubber surface will soften and will have to be re-surfaced. This could be done in the in-house workshop. Spares are recommended to minimize downtime.

Since embossing rollers are very expensive, special attention is needed to prevent surface damage or retention of any tiny particles from the hot coating surfaces. After cleaning the rollers, the use of adequate covers would be useful.

10.2.2 Lean manufacturing practice

Lean practices basically mean elimination of *waste* and in a manufacturing atmosphere primarily materials waste and unnecessary efforts will take precedence. In an organization for overall effective lean practices, it should start right from the top, down the line and probably ending in shipping. If each department is asked to do it, it is unlikely to end in good results as seldom will they identify their own shortcomings. The ideal would be (a) a special management team, (b) another department team or (c) a professional body from outside which would be the best.

'Lean processing' or 'lean production' is a production practice that considers the application of resources for any operation, other than the creation of 'value' for a product or service, to be wasteful and thus a target for elimination. Elimination of waste during each process cycle results in increase of productivity and collectively contributes to wards increased profitability. From the perspective of a customer who 'consumes' a product or service and is paying for this 'value' and is willing to pay for it, if the quality and price is right. One might say that lean processing is centred on an increased value at less cost with quality always in mind.

Lean practices can be applied to any business or industry and if applied effectively can give great surprises and show new possibilities. The methods of waste reduction are both exciting, challenging and rewarding. General lean methodology in a manufacturing operation is based on three types of waste reductions: no value-added work, overburdening a process and uneven process flow.

A lean operation is a variation on the theme of efficiency and is based on optimizing smooth flow. In a manufacturing operation, it would be increasing the efficiency of a process and good coordination between processes with decreasing of wastes of all processes. It could be also termed 'fine-tuning'.

10.2.3 Quality control system in brief

Quality control is a vital factor in any industrial operation. For plastics processing plants such as injection moulding, blow moulding or extrusion a QC system would be more or less easy after selecting an ISO standard like 9001, 9002 or QS 9000 and implementing a statistical processing control system. However, for the manufacture of artificial leather being a coating process it will be different.

From experience, the author would like to recommend a variation of the normal QC procedures for more effective control of quality. First, naturally, an organization would decide on an international standard like British Standard Specifications (BSS), American Standards for Testing Materials or DIN (German) or other. The second step would be to choose an ISO standard and then implement a QC system to meet these standards. Unlike in processing plastics, there are many variants in this manufacture and the author recommends that QC systems be implemented on the basis of each processing sector, a system to suit each sector rather than on an overall basis.

The author recommends that the overall operation for an artificial leather manufacturing plant be divided into the following sectors:

a. **Raw materials**

Will mainly consist of chemicals and the fabrics to be coated. The chemicals will arrive on a production floor with certification by the suppliers and very large manufacturers will check the quality of each using small samples in an in-house lab. However, this will need high skills and for the average manufacturer the standard certifications and guidelines provided should suffice to accept them as good quality.

The other bulk material will be the silicone-coated release paper containing embossed patterns on which the first coating (surface layer) will be done. These are huge rolls, very wide and heavy and cannot be handled manually. Here again, a manufacturer will have to accept the specifications as provided by reputed suppliers.

b. **Fabrics**

Since different types of fabrics like cotton, synthetics, bamboo cloth (new concept) or others, both woven and knitted can be used, and special attention is needed in checking the quality. Whatever international standard is used, the weight per square yard per metre of the fabric should be as specified for each class of artificial leather. For example, BSS may specify a weight of 4 oz. per square yard for direct coated leather made with woven cotton cloth. This will be a factor when later the finished coated cloth is tested for quality especially for tear strength, tensile strength and so on.

In woven cloths, the overall width becomes important for consistency in width and should be free of wrinkles, knots, fibres sticking out and other deficiencies that would hinder a smooth coating. General widths for coated cloths could be 54, 72 or 108 in and to achieve these final widths, the base cloths should be 1.5–2.0 in wider. For example, to get a final coated width of 54 in, the starting width should be at least 56 in to counter the warp/weft shrinkage due to the tension on the coating machine. The cloths should be free of joints as these will pose coating problems as well as short lengths.

The checks on knitted fabrics for indirect coating are easier as there is no shrinkage of width since the fabric will be 'laid down' on the last foam coating and the weight weave becomes more important. If the 'knit' is too open, the liquid foam coat will seep through the fabric when pressed going through the laminator.

The following quality control charts can be used as appropriate to each of the above processing sectors.

10.2.3.1 What are control charts?

A control chart is a graphical representation of the quality of a particular feature of a product being made, for example, weight, density and dimensions generated by a process. These charts will have an upper control limit and a lower control limit and a median related to the acceptable pre-set tolerances. The frequency of checking and recording may vary with each different product.

Why use a control chart? The periodic readings taken and recorded while a process is in progress will show 'how well' it is performing to the pre-set tolerances and what action is needed, if necessary, to keep the process within limits.

10.2.3.2 X–R control chart

An X–R chart is a chart used for recording variable data.
    For example:
- Measuring a feature of a product
- Quantitative (length, density, size or weight)

These charts will have four basic sections. Section one will show
- Name of customer or product code
- Product identity
- What feature is being measured
- How often it is done

Section two will contain the graph or grid with limits, while section three will show the following data:
- Date
- Time
- Measured values
- Sum total

- Averages
- Range
- Operator initials

Section four is the graph and readings for calculating and recording the range.

10.2.3.3 P-chart

A P-chart is used for recording attribute data. This documentation is based on determining the good or bad, accept or reject concept of a product/products. In the case
of manufacturing artificial leather, a P-chart application would be appropriate for
checking fabrics and especially at the final stage of packing. Here, the packers in
addition to the standard label of date, time, product ID/code or others could use a
colour code like green (accept), yellow (on hold) and red (reject).

Artificial leather may be packed in roles of 50 or 100 m where they can be manually handled. Unlike in other manufactures, rejects/wastes could be minimized by
re-packing the yellows and even the reds into smaller packs (short lengths) which
would have a market among the smaller end-users like small volume shoe manufacturers, upholsterers and others. Here, the P-chart data will be very helpful in weed-
ing out what actually should be rejected.

10.3 Parameters for monitoring efficiency

Generally, an organization would record processing data on a daily basis but it is
likely that these will reach management every month or even on a quarterly basis or
other in the form of reports. By this time it may be too late as losses would have
already occurred which cannot be reversed.

However, in addition to processing drawbacks like downtime, absenteeism,
shortage of raw materials, accidents and so on, there are some parameters that management can use to monitor how well a process or overall process is performing.
These will enable corrective action where necessary.

10.3.1 Cost per kilogramme

Total gross costs of operation versus cost of raw materials consumed during a period of time.

For example, total costs = ex-factory + administration + marketing + other =
$72,000/-

Raw materials consumed during this same period = 1,200 kg.
Therefore, cost per kilo = ($72,000 ÷ 1,200) = $60.00 per kg.

Compare with pre-set standards.

Note: this calculation can also be done on the basis of ex-factory only.

10.3.2 Return per yard/metre

The return per yard of finished artificial leather would be the total gross value of sales and saleable leather versus total costs of manufacture for a given period of time.

For example, 1,000 yards of finished leather @ $20 per yard = $20,000

Total costs of manufacture for the same period of time = $10,000

Therefore, the return per yard = ($20,000 − $10,000) = $10,000 ÷ 1,000 = $10 per yard

Compare with pre-set standards.

10.3.2.1 'Kick-backs' from wastes

In the manufacture of artificial leather, there are two avenues of 'kick-back' revenues: from sales of coating material wastes and short lengths. It is virtually impossible to make the exact amount of coating material needed and a manufacturer will ensure that a production run will have sufficient coating material, even if some of it will be left over. These should be kept separately and will be ideal to produce some artificial leather having a psychedelic coloured surface. The other will be from sales of short lengths of finished products that can be sold to small users, for example, shoe manufacturers.

10.3.3 Contribution margins

In general, artificial leather manufacturing operations are based on a wide range of types, colours, surface designs, thicknesses, surface finishes (gloss/matt), solid coated or foamed and so on. In a multiple product manufacturing operation in order to maximize profits, it may be essential to decide which products are viable and which should be eliminated or taken off the production line.

This will have to be information from at least three sectors like production, accounts and sales analysis and a final decision by management. The tool to do this will be the 'contribution factor' of each product. A contribution factor/margin is calculated by the amount of money a product generates towards meeting the overall overheads and calculated as: unit sale price − variable costs of that product.

In decision-making, it should be remembered that some products may have large sales volumes with small margins while some others may have smaller sales volumes but with large contribution factors. Other factors like popularity, essential feeder material for industries or others may also influence decision-making.

10.3.4 Breakeven point

A breakeven analysis will indicate how much sales volume is needed to meet total expenses and all sales beyond this point being considered as profit. This can be calculated for individual products or on an overall basis. This information is especially useful in product pricing policy and setting up targets. This calculation can be done by formula or represented as a graph.

Use the formula: BEP = TFC/USP – VCUP

where BEP is breakeven point (no. of units), TFC is total fixed costs, USP is unit selling price and VCUP is unit variable cost.

10.3.5 Pricing for exports

Artificial leather manufacturers have high potential for exports of their products. Some may export directly to a buyer or go through an agency or other. The price structures for local sales and exports are different. While local prices are probably decided by a marketing division of an organization, exports prices need more attention with the need to keep in mind other suppliers. As a general rule, export prices are worked out on the basis of ex-factory cost +10% to 15%. This would also depend on the volume of the order, and a manufacturer may decide on a minimal markup on very large orders.

Bibliography

[1]   Defonseka, Chris. "Practical Guide to Flexible Polyurethane Foams." Smithers Rapra UK, 2013.

Appendix A
List of some machinery and equipment suppliers

1. HK Zion Industry Co. Ltd. – Guangdong China
2. Qundao Huarui Jiahe Machinery Co. Ltd. – Shandong China
3. Artisan Engineering Works – Shahari India
4. Crowm Machinery Co. – Taiwan
5. Shine Kon Enterprises Ltd. – Taiwan
6. Taiwan Forward Machinery Co. – Taiwan
7. Forma Corporation Inc. – USA
8. Torrey Hills Technologies – USA (mixing equipment)
9. Charles Ross & Son Company – USA (mixing and blending equipment)
10. Crown Machinery Co. –Taiwan (embossing machines)

Full coating line machinery suppliers will generally supply mixing and blending equipment and also standard embossing units. If specialty embossing patterns and finishes are desired, these equipment can be purchased separately as custom-made.

https://doi.org/10.1515/9783110716542-011

Appendix B
List of raw material suppliers

Table in this appendix shows some of the popular manufacturers and suppliers of polymers. They will be able to advise processors about good sources for additives, dyes, pigments and also others. Polymers are generally available as powders, liquids, granules or pellets, either natural in colour or self-coloured. Basic packs include 25 kg paper bags or in larger 400–500 lb bulk packs, except for the polymers in liquid form. Countries shown are the main manufacturing sources but their products will be available from their agents/distributors in many countries.

Supplier	Country	Products
Dow Corporation	USA	All
BASF	Germany	All-speciality products
Bayer AG	Germany	All-speciality (PUR)
Huntsman Corporation	Europe	Speciality PUR
Chemcontrol Limited	USA	PUR
Issac Industries Inc.	USA	PUR
ERA Polymers Ltd.	Australia	Two-component PUR systems
Bio-based Technologies LLC	USA	Speciality polymers
Union Carbide Limited	Canada	Polymers
Chemeon Surface Technology	USA	Dyes and pigments
Elkem Company	USA	Silicones
Stahl Company	USA	Coatings for fabrics

PUR, polyurethane.

This table, as compiled by the author, highlights a few major suppliers. Materials can be purchased as separate components and mixed and blended on the production floor or purchased as coating systems.

https://doi.org/10.1515/9783110716542-012

.

Appendix C
List of suppliers of biomass fillers and stiffening agents

The following table shows some randomly selected suppliers of biomass fillers and stiffening agents. Since most are manufacturers, customized products can be purchased. Best sources for finding suitable suppliers for additives, dyes, pigments and others are these biomass suppliers who can also recommend the best products to use.

Supplier	Country	Products
Composite Materials Co. Inc. CT	USA	Wood flour, walnut shell flour, rice hull flour, sisal, corncob flour
Hammond Roto Finish -MT	USA	Rice hull flour
Mid-Link International Co. Ltd. European Office/Shanghai China	Germany	Rice hull ash
Silicon India	India	Rice hull ash, powder, pellets
NK Enterprises	India	Rice hull ash
ADF Asset & Investments	UK	Wheat hulls
Agrilectric Research	USA	Rice hull ash
Tianjin Glory Tang Co. Ltd.	China	Bamboo fibre
Siddhi Vinayak Enterprises	India	Bamboo fibre
Zenco Global Enterprises	Malaysia	Soya bean flour, wheat hulls
MM Chemical India	India	Composite polymer powders, high-density polyethylene, low-density polyethylene, polypropylene, ethylene vinyl acetate, customized powders
Kanju Industrial (HK) Ltd.	China	Graphite powder for polymers
Rice Hulls Speciality Inc.	USA	Rice hulls, rice hull powder

This table, as compiled by the author, shows some key suppliers but many more are available covering a wider range of products.

https://doi.org/10.1515/9783110716542-013

Polymer glossary

Amorphous having no ordered arrangement. Polymers are amorphous when their chains are tangled up in any old way. Polymers are *not* amorphous when their chains are lined up in ordered crystals.

Anion an atom or molecule which has a negative electrical charge

Cation an atom or molecule which has a positive electrical charge

Complex two or more molecules which are associated together by some type of interaction of electrons, other than a covalent bond

Copolymer a polymer made from more than one kind of monomer

Covalent bond a joining of two atoms when the two share a pair of electrons

Crosslinking crosslinking is when individual polymer chains are linked together by covalent bonds to form one giant molecule

Crystal a mass of molecules arranged in a neat and orderly fashion. In a polymer crystal, the chains are lined up neatly like new pencils in a package. They are also bound together tightly by secondary interactions.

Elastomer rubber. Hot shot scientists say a rubber or elastomer is any material that can be stretched many times its original length without breaking *and* will snap back to its original size when it is released.

Electrolyte a molecule that separates into a cation and an anion when it is dissolved in a solvent, usually water. For example, salt, NaCl separates into Na^+ and Cl^- in water.

Elongation how long a sample is stretched when it is pulled. Elongation is usually expressed as the length after stretching divided by the original length.

Emulsion a mixture in which two immiscible substances, like oil and water, stay mixed together, thanks to a third substance called an *emulsifier*. The emulsifier is usually something like a soap, whose molecules have a water-soluble end and an organic-soluble end. The soap molecules form little balls called *micelles*, in which the water-soluble ends point out into the water, and the organic-soluble ends point into the inside of the ball. The oil is stabilized in the water by hiding in the centre of the micelle. Thus, the water and oil stay mixed.

Entropy disorder. Entropy is a measure of the disorder of a system.

First-order transition a thermal transition that involves both a latent heat and a change in the heat capacity of the material

Free radical an atom or molecule which has at least one electron which is not paired with another electron

Gel a crosslinked polymer which has absorbed a large amount of solvent. Crosslinked polymers usually swell a good deal when they absorb solvents.

Gem diol a diol in which both hydroxy groups are on the same carbon. Gem diols are unstable. Why are they called *gem* diols? It is short for *geminal*, which means 'twins'. It is related to the word *gemini*.

https://doi.org/10.1515/9783110716542-014

Glass transition temperature the temperature at which a polymer changes from hard and brittle to soft and pliable

Heat capacity the amount of heat it takes to raise the temperature of one gram of a material by 1 °C.

Hydrodynamic volume the volume of a polymer coil when it is in solution. This can vary for a polymer depending on how well it interacts with the solvent, and the polymer's molecular weight.

Hydrogen bond a very strong attraction between a hydrogen atom which is attached to an electronegative atom, and an electronegative atom which is usually on another molecule. For example, the hydrogen atoms on one water molecule are very strongly attracted to the oxygen atoms on another water molecule.

Ion an atom or molecule which has a positive or a negative electrical charge.

Latent heat the heat given off or absorbed when a material melts or freezes, or boils or condenses. For example, when ice is heated, once the temperature reaches 0 °C, its temperature will not increase until all the ice is melted. The ice has to absorb heat in order to melt. But even though it is absorbing heat, its temperature stays the same until all the ice has melted. The heat required to melt the ice is called the *latent* heat. The water will give off the same amount of latent heat when you freeze it.

Le Chatlier's principle this principle states that if a system is placed under stress, it will act so as to relieve the stress. Applied to chemical reactions, it means that if a product or byproduct is removed from the system, the equilibrium will be upset, and the reaction will produce more products to make up for the loss. In polymerizations, this trick is used to make polymerization reactions reach high conversion.

Ligand an atom or group of atoms which is associated with a metal atom in a complex. Ligands may be neutral or they may be ions.

Living polymerization a polymerization reaction in which there is no termination, and the polymer chains continue to grow as long as there are monomer molecules to add to the growing chain

Matrix in a fibre-reinforced composite; the matrix is the material in which the fibre is embedded, the material that the fibre reinforces. It comes from a Latin word which means 'mother', interestingly enough.

Modulus the ability of a sample of a material to resist deformation. Modulus is usually expressed as the ratio of stress exerted on the sample to the amount of deformation. For example, tensile modulus is the ration of stress applied to the elongation which results from the stress.

Monomer a small molecule which may react chemically to link together with other molecules of the same type to form a large molecule called a polymer

Olefin metathesis a reaction between two molecules, both containing carbon–carbon double bonds. In olefin metathesis, the double bond carbon atoms change partners to create two new molecules, both containing carbon–carbon double bonds.

Oligomer a polymer whose molecular weight is too low to really be considered a polymer. Oligomers have molecular weights in the hundreds, but polymers have molecular weights in the thousands or higher.

Plasticizer a small molecule that is added to polymer to lower its glass transition temperature

Random coil the shape of a polymer molecule when it is in solution, and it is all tangled up in itself, instead of being stretched out in a line. The random coil only forms when the intermolecular forces between the polymer and the solvent are equal to the forces between the solvent molecules themselves and the forces between polymer chain segments.

Ring-opening polymerization a polymerization in which cyclic monomer is converted into a polymer which does not contain rings. The monomer rings are opened up and stretched out in the polymer chain.

Secondary interaction interaction between two atoms or molecules other than a covalent bond. Secondary interactions include hydrogen bonding, ionic interaction and dispersion forces.

Second-order transition a thermal transition that involves a change in heat capacity, but does not have a latent heat. The glass transition is a second-order transition

Soap a molecule in which one end is polar and water-soluble and the other end is non-polar and organic-soluble, such as sodium lauryl sulfate. These form micelles in water, little balls in which the polar ends of the molecules point out into the water, and the non-polar ends point inward, away from the water. Water-insoluble dirt can hide inside the micelle, so soapy water washes away dirt that plain water cannot.

Strain the amount of deformation a sample undergoes when one puts it under stress. Strain can be elongation, bending, compression or any other type of deformation.

Strength the amount of stress an object can receive before it breaks.

Stress the amount of force exerted on an object, divided by the cross-sectional area of the object. The cross-sectional area is the area of a cross section of the object, in a plane perpendicular to the direction of the force. Stress is usually expressed in units of force divided by area, such as N/cm^2.

Termination in a chain growth polymerization, it is the reaction which causes the growing chain to stop growing. Termination reactions are reactions in which none of the products may react to make a polymer grow.

Thermoplastic a material that can be moulded and shaped when it is heated.

Thermal transition a change that takes place in a material when you heat it or cool it, such as melting, crystallization or the glass transition.

Thermoset a hard and stiff crosslinked material. Thermosets are different from *thermoplastics*, which become mouldable when heated. Thermosets are crosslinked, so they do not. Also, they are different from crosslinked *elastomers*. Thermosets are stiff and do not stretch the way elastomers do.

Toughness a measure of the ability of a sample to absorb mechanical energy without breaking, usually defined as the area underneath a stress-strain curve.

Transesterification a reaction between an ester and an alcohol in which the -O-R of the ester and the -O-R' group of the alcohol trade to change places.

Some terms in artificial leather industry

(Compiled by the author)

Artificial	imitation
Air bubbles	generated by shear forces during mixing
ASTM	American Standard Testing Materials
Bias	diagonal direction on woven fabric 45° warp to weft
Bio-fabrics	fabrics made from fibres of pineapple, bamboo or other
Count	size of thread (relation between length and weight)
Coating	protective or for aesthetic value
Coating angle	coating knife angle for smooth/uniform application of coating
Catalyst	chemical used to promote or slowdown a chemical reaction
Copolymer	used for PVC coatings
Doctor-knife	a wide blade used for coatings, which can have different configurations
Denier	unit of measurement of density/weight related to a fixed length of thread
Draft	diagram for setting up a weave pattern
Dams	adjustable plates attached to wide coating knife at both ends
Density	weight per unit volume
Exothermic	giving off heat
Endothermic	absorbing heat
Emulsion	particles suspended in a solution
EPI	ends per inch (number of warp threads per inch of a woven fabric)
Edges	borders on either side of a fabric across the weft
Face	right side of a textile or weave
Filler	component used for reducing costs
Fabric	general term for any manufactured cloth
Fibre	thread-like tissue which can be natural, synthetic or biomass
Fluid	mass in a liquid state
Gas loss	weight of chemicals lost during foaming
Knitted fabrics	used in indirect or transfer polymeric coatings
Laminate	many types, for example, knitted fabric on polymeric coating
Lubricant	component used for increasing fluidity
Pile	supplementary threads projecting from a ground fabric
PPI	picks per inch. Number of weft threads per inch of woven fabric
Pilling	formation of small balls of fibres called pills on the surface of a fabric
Porous	open weave, where a material will seep through
Plasticizer	component used for softening a polymer
Release paper	plain or embossed silicone-coated paper
Skin coat	generally refers to the final or top coat
Stabilizer	component used to prevent degradation due to heat
Synthetic yarn	fibres made from synthetic materials like nylon, polyester and polyethylene
Spatula	tool used for spreading polymeric mix on fabric at coating head
Thickening agent	component used to prevent polymeric mix from seeping through fabric
Thread count	number of warp ends or picks per unit of measure
Toxic gases	gases generated especially during coating solvent-based polyurethanes

https://doi.org/10.1515/9783110716542-015

Triple-roll mill	a three-roll mill for making pastes, for example, filler/pigment
Viscoelastic	foams having four-dimensional properties
Yarn	threads used for weaving or knitting
Warp	lengthwise in a fabric
Weft	widthwise in a fabric
Woven fabrics	used in direct polymeric coatings

www.ingramcontent.com/pod-product-compliance
Lightning Source LLC
Chambersburg PA
CBHW081539220326
41598CB00036B/6488